The Guide for the Disabled Traveller
ON THE MOVE

THE SERVICE FOR DISABLED MOTORISTS

Published by RAC Publishing, RAC House,
Bartlett Street, South Croydon CR2 6XW

© RAC Motoring Services Ltd 1995

This book is sold subject to the condition that it shall not, by way of trade, or otherwise, be lent, re-sold, hired out or otherwise circulated without the publisher's prior consent in any form of binding, or cover other than that in which it is published.

All rights reserved. No parts of this work may be reproduced, stored in a retrieval system or transmitted by any means without permission. Whilst every effort has been made to ensure that the information contained in this publication is accurate and up-to-date, the publisher does not accept any responsibility for any error, omission or misrepresentation. All liability for loss, disappointment, negligence or other damage caused by reliance on the information contained in this guide, or in the event of bankruptcy, or liquidation, or cessation of trade of any company, individual or firm mentioned is hereby excluded.

ISBN 0 86211 322 9

A CIP record for this book is available from the British Library

Consultant Editor: Arthur Ledgard
Editor: John Andrews
Design: Chuck Goodwin

Advertising Managers: West One Publishing
Portland House, 4 Great Portland Street, London W1N 5AA
Tel: 0171-580 6886

Printed and bound by BPC Paulton Books Ltd., Great Britain

Contents

RAC SERVICES

RESPONSE	6
RAC Service for Disabled Motorists	6
ASSOCIATED CLUBS	7
LEGAL SERVICES	7
TECHNICAL SERVICES	7
Technical Advice	7
Vehicle Examination Service	8
INFORMATION SERVICES	8
Traffic and Travel Information for the UK and Europe	8
Premium Rate Information Service	8
RAC Routes Service	8
How to Get Your Personal Route Pack	8
Autoroute Express – The Intelligent Road Atlas	8
TRAVEL SERVICES	9
RAC Eurocover Insurance	9
PUBLISHING	9
HOTEL SERVICES	9
INSURANCE SERVICES	10
RAC Motor Insurance	10
RAC Home Contents Insurance	10
Caravan and Small Craft Insurance	10
SIGNS SERVICE	10
PRIVILEGES AND DISCOUNTS	10
Autoglass	10
Beaulieu National Motor Museum	10
Bupa	10
EuroDollar	11
Hometune	11
National Tyres and Autocare	11
Superdrive	11
PROTECTING THE INTERESTS OF THE MOTORIST	11
Public Affairs	11

DRIVING AND THE DISABLED

THE ORANGE BADGE SCHEME	12
You can get a Badge if...	12
Where and How to Apply	12
Where the Scheme does not Apply	12
Where to Park	12
Red Routes	13
Where not to Park	13
How to Use the Badge	13
How to use the Special Parking Disc	13
Giving a Vehicle Registration Number	13
Your Duties as a Badge Holder	13
Other Benefits for Badge Holders	14

MOTABILITY	14
INDEPENDENT DRIVING ASSESSMENT CENTRES	15
Banstead Mobility Centre	15
Disability Action	15
Mobility Advice and Vehicle Information Service (MAVIS)	15
Mobility Information Service (MIS)	16
Wales Disabled Driving Assessment Information Centre	16
Regional Driving Assessment Centres	16
CAR MANUFACTURERS OFFERING DISCOUNTS AND CONCESSIONS FOR DISABLED DRIVERS	17
FIRMS THAT MANUFACTURE OR CONVERT VEHICLES	19
SHOPMOBILITY SCHEMES	26
The National Federation of Shopmobility (NFS)	26
FACILITIES AT SUPERMARKETS	31
THE ROYAL ASSOCIATION FOR DISABILITY AND REHABILITATION (RADAR)	32
NATIONAL KEY SCHEME (NKS)	33
DIAL UK	33
ROUTE FINDING AND ACCESS MAPS PROJECT (RAMP)	35

TRAVEL AT HOME AND ABROAD

MOTORWAY SERVICE AREAS	34
Motorway and A-road Accommodation	37
TOLL CONCESSIONS	38
PROVISION FOR DISABLED TRAVELLERS ON CAR FERRIES	39
THE CHANNEL TUNNEL	45
HIRING A CAR AT HOME AND ABROAD	46
UK AIRPORTS	46
RECIPROCAL PARKING ARRANGEMENTS IN EUROPE FOR ORANGE BADGE HOLDERS	51
USEFUL ADDRESSES	54
Organisations	54
Holiday Services and Accommodation	58
Useful Products for the Disabled	59
USEFUL PUBLICATIONS	60

PLACES TO VISIT 66

PLACES TO STAY 74

in ACTIVITY

Looking for a fun packed adventure holiday, a lazy break in the country or a little bit of both?

At the Kielder Calvert Trust Centre in the heart of beautiful Northumberland...

THE CHOICE IS YOURS

- **SAILING**
- **CLIMBING**
- **SWIMMING**
- **CANOEING**
- **RIDING**
- **OR SIMPLY ENJOYING!**

Purpose built for those of our guests with disabilities, the centre offers a varied menu of courses and activities, run by qualified staff who are constantly on hand to offer support and encouragement.

For those guests who prefer a more leisurely holiday the lakeside centre, with its own heated swimming pool, is perfect for enjoying those idle moments or as a base from which to explore this spectacular part of the Northumberland countryside.

Activity or *in* ACTIVITY?... the Choice really is Yours!

For more holiday details contact:
Calvert Trust, Kielder Water, Hexham, Northumberland, NE48 1BS
Telephone: 01434 250232   Fax: 01434 250015

Introduction

The Royal Automobile Club was founded almost 100 years ago to encourage motoring and protect the interests of the motorist. The Club enjoys Royal Patronage and its President HRH Prince Michael of Kent is an active participant in its affairs.

The aim of the RAC is to provide a balanced portfolio of services which enhance the security and the peace of mind of motorists, and to champion socially responsible developments in motoring matters, anticipating the needs of all motorists and providing an independent voice for them. It was therefore appropriate that, with the growing number of disabled motorists able to take advantage of the independent mobility offered by the motor car, the RAC introduced 'Response' – a service specifically geared to disabled motorists. By combining the RAC's wealth of experience in delivering its services in the most responsive, relevant and friendly way with world-leading technology, RAC Response is able to offer disabled motorists who have broken down a genuinely flexible solution, tailored to their individual needs.

On The Move - The RAC Guide for the Disabled Traveller is designed to provide information on all aspects of motoring for people with disabilities and those who drive with disabled passengers. It includes details for the services available to Members of RAC Response together with a wealth of useful information for disabled drivers at home and abroad. It gives details of centres which can offer advice and assessment on the requirements for disabled drivers; details of accesssible places to visit and of RAC hotels which welcome disabled travellers, plus a great deal of other useful information. We are sure that it will provide an answer to many of the questions disabled motorists ask.

Jeffrey Rose CBE
Chairman

RAC Services

RAC RESPONSE

■ RAC SERVICE FOR DISABLED MOTORISTS

RAC Response is the first service to be offered by a motoring organisation specifically for disabled people and is open to all holders of a local authority-issued Orange Badge. The latest technology combines with the practical skills and experience of our highly trained Rescue Services staff to offer a complete, 24-hour, all-year, nationwide roadside service.

As a member of RAC Response you will be aware of the additional problems that a disabled motorist can face with a breakdown. There may be urgent needs other than attention to the car. You can be assured as an RAC Response Member that if your journey is interrupted, the RAC, in addition to looking after your car, will ensure that you receive whatever assistance is necessary to enable you to reach your destination as quickly as possible and with the minimum fuss and bother.

HOW TO CONTACT US

If you are able to leave the vehicle, you can contact us on our emergency free number – 0800 82 82 82. Deaf drivers can contact the RAC Rescue Services through our Minicom Supertel unit using a personal printer and modem on 0800 62 63 89.

If you are unable to leave the vehicle, then attract assistance by turning your hazard warning lights on and hanging the RAC 'help' pennant from your car window. When passing information on the breakdown to a third party to relay to the RAC, complete one of the sheets in the 'breakdown details' pad supplied with your membership pack.

YOUR PERSONAL INCIDENT MANAGER

When you give your RAC Response membership number, the person answering your call will know that you may have special requirements. If it is appropriate, you will be put in direct contact with a Personal Incident Manager, who is trained to listen to your requirements and translate them into a practical tailor-made solution – quickly and effectively.

Even if you do not need to make direct contact with a Personal Incident Manager during the initial call, one will be monitoring the progress of the response to your call, and if at any point during the incident you do need to talk to a Personal Incident Manager, one will be on hand.

In addition to monitoring your situation and, where required, assisting in making any special arrangements, your Personal Incident Manager will also pass on any messages to family and friends, or even the emergency services if necessary.

Rescue, Recovery, 'At Home'

Membership of RAC Response automatically entitles you to the traditional RAC services of Rescue, Recovery and 'At Home', and includes family-based cover, so that anyone – family, friends or colleagues – driving your nominated car with your permission can call on RAC services. You and your spouse also benefit from personal cover, which gives RAC protection when you travel in any vehicle, including a hired car or one belonging to a friend.

It is usually possible to repair your vehicle at the roadside, but if not, your Personal Incident Manager will, if you wish, look into the possibility of further repairs being carried out at the nearest RAC-approved garage. We will also come to your aid if vandals or thieves immobilise your car.

If you have a serious breakdown or accident and your vehicle cannot be repaired within a reasonable time at the roadside or at a nearby garage, we will take you, up to four passengers and your car straight home or to a single destination of your choice, anywhere in the UK. You do not necessarily have to travel with your vehicle – we can recover it while you continue your journey by other means. This includes journeys to and from Northern Ireland, but excludes the cost of ferry tickets for vehicle and passengers. The service also covers caravans and trailers. It can even cover you if you are unable to drive through serious injury or illness.

The inclusion of the 'At Home' service means that if your car breaks down at or near home, one free call will bring assistance from the RAC. If we cannot repair your vehicle on the spot, we will arrange for it to be towed to a nearby garage, and if you need any message to be passed on, the RAC will be happy to do this for you.

Assured Mobility

If your car does need to be towed to a garage or to your home, you may still wish to continue your journey. When this is the case you will be put in touch with your Personal Incident Manager, with whom you will be able to decide on the most appropriate means of onward transport, taking into account any special requirements you have. Train, hire car or taxi are all possibilities. The RAC will obtain an automatic hire car in some circumstances, subject to availability. However, we are not able to provide specially adapted vehicles. The one thing you can be sure of is that we will get you to your required destination.

If your car is stolen

We will also help if your vehicle is stolen. First, report the theft to the police and get an incident number from them. Give this number to us and we will help you get home. Your Personal Incident Manager can arrange for a replacement car for you for up to three days – you only have to pay the petrol – or you and your passengers can complete your journey by rail. If you prefer to make your own arrangements for getting home – by coach, taxi or even plane – the RAC will reimburse the cost up to a generous limit.

Once your vehicle has been found, we will be happy to arrange for its safe recovery and return to your home. The service does not include taking you to the car or storing the vehicle. If you do not use the additional services when your car is stolen, they can be used when the vehicle is found.

RAC Associated Clubs

The RAC has a long-established relationship with clubs representing a wide range of motoring interests, through the RAC Associated Clubs scheme. The clubs play an important constitutional role within the RAC through their representation at the RAC General Council and on the RAC Associate Committee.

Individual Members can also benefit from the association between their club and the RAC. There are preferential rates on certain RAC services, and club funds profit also, as a commission is received on RAC membership taken out through the club schemes.

Those organisations which specifically promote the interests of the disabled motorist on a national basis, and have Associated Club status, are

The Disabled Drivers' Association (DDA)
Ashwellthorpe, Norwich, Norfolk NR16 1EX
Tel: (01508) 489449

Disabled Drivers' Motor Club (DDMC)
Cottingham Way, Thrapston, Northamptonshire
NN14 4PL Tel: (01832) 734724

If you would like further information on either of these organisations, please contact them direct. If you would like information on any of the other clubs with Associated status, please contact the RAC Associated Clubs Section on 0345 41 41 51, Monday to Friday 9am to 5pm.

RAC Legal Services

Expert free legal advice is available to RAC Members by calling 0345 300 400, Monday to Friday 9am to 5pm.

If you have been involved in an accident, the RAC Legal Department can help you. We have an accident claims service with a nationwide network of specially selected solicitors of proven expertise. In these cases negotiation and, if necessary, litigation are provided upon payment of a modest registration fee.

For Members who have a dispute with a garage or retailer, we can also provide a formal negotiation (not litigation) service through our expert Advisors. Again, this service is subject to the payment of a registration fee.

RAC Legal Services was established in 1908 and has acquired a reputation for excellence through more than 80 years' experience of acting on behalf of Members. To take advantage of this valuable benefit call the Legal Department on 0345 300 400.

NOTE: The RAC reserves the right to refuse legal representation in certain cases. Though we can give advice, it is RAC policy not to provide representation in drink- or drug-related cases.

RAC Technical Services

■ TECHNICAL ADVICE

As an RAC Response Member you can receive technical advice from our engineers, who can make sure that your problems have been handled fairly and professionally. Advice is offered on the price, running costs, repair and maintenance of vehicles. Our engineers will also check invoices and, for a small fee, will negotiate in cases of dispute with a garage or manufacturer to ensure that you get a fair deal. Call the RAC National Technical Centre on 0345 345 500, Monday to Friday 8.30am to 5.30pm.

Environmental enquiries

The environmental effects of motoring are of special concern to the RAC. Call us on 0345 345 500 if you have any questions about the effects of

RAC SERVICES

changing to unleaded fuel, or whether your car can use it. Our engineers will provide information on the advantages of catalytic converters or other 'green' issues relating to motoring.

■ VEHICLE EXAMINATION SERVICE

Inspections through the RAC's Vehicle Inspection Service cost a fraction of the value of a car, are easy to arrange and take the uncertainty out of buying a used car. To book a vehicle examination or for more information on the service call free on 0800 333 660, Monday to Friday 8.30am to 6pm, Saturday 9am to 5pm.

RAC INFORMATION SERVICES

■ TRAFFIC AND TRAVEL INFORMATION FOR THE UK AND EUROPE

The RAC has an enormous amount of expertise on motoring and travel. By calling 0345 333 222 (calls charged at local rate) you will access the RAC's comprehensive telephone information service which operates Monday to Friday 7am to 7pm, Saturday 9am to 5pm and Sunday 9am to 7pm.

■ PREMIUM RATE INFORMATION SERVICE

Out of operating hours the following information services are available:
- UK motorist's hotline. The latest traffic conditions for London, UK motorways and major routes Tel: 0891 500 242
- traffic and roadworks in Europe Tel: 0891 500 241
- European touring information, documentation and motoring regulations Tel: 0891 500 243
- ferryline details of delays, cancellations and sea conditions on routes from the major ports, plus road traffic reports on problems that may affect your journey to a port. Tel: 0891 700 312.

Calls are charged at 39p per minute cheap rate and 49p per minute at all other times.

■ RAC ROUTES SERVICE

For a small charge the RAC will prepare a route tailored to your individual requirements. UK and Irish routes contain junction-by-junction driving instructions, including motorway service facilities, town plans, RAC roadworks report, a detailed map of the country and other vital information.

For journeys in mainland Europe we have three types of route pack:

EUROPEAN ROUTE AND TRAVEL PACK

Everything you need to reach your destination – a route with junction-by-junction driving instructions, town plans, RAC roadworks report and maps of the countries you plan to visit, plus a fact-filled motoring guide for each country.

UK AND EUROPEAN ROUTE AND TRAVEL PACK

Our most comprehensive service, this includes not only the material in the European pack but also the best route from your home town to your port of departure, UK town and port plans, UK roadworks report and UK road map.

EUROPEAN ROUTE AND TOWN PLANS

You might be familiar with the driving conditions and regulations in Europe, but you would undoubtedly benefit by applying for an RAC computerised route with town plans tailored to your specific needs.

■ HOW TO GET YOUR PERSONAL ROUTE PACK

Call 0345 333 222 and our experts will help you on all details and charges. To help us give you the best service please have your credit card number and your specific route request when you call. Allow 14 days for us to prepare your route and send it back to you by post, or opt for our 48-hour delivery service available for a supplement. Please quote A1 in all calls and correspondence. You can find a route application form at any RAC Travel Centre or in the RAC Travel Services brochure.

■ AUTOROUTE EXPRESS – THE INTELLIGENT ROAD ATLAS

For those of you owning or having access to a PC, we bring you AutoRoute Express – route-planning software that saves up to 20% on driving time and costs, and is so easy to use.

Available in Windows, DOS and Macintosh versions, and with clear maps and straightforward directions that can be printed out or viewed on screen, AutoRoute Express is the ideal tool for both business travel and holiday plans.

RAC Members qualify for a 25% discount when purchasing AutoRoute Express through RAC Information Services on 0345 333 222 (calls charged at local rate). Retail prices start from just £99.95 (including VAT) for the DOS version.

RAC Travel Services

RAC Travel Services is a 'one stop' travel shop providing a comprehensive range of products and services for holiday and business travellers to Europe. Call free on 0800 550 055 (Monday to Friday 8.30am to 6pm, Saturday 9am to 5pm) for full details. Here are just some of the ways in which we can help.

■ RAC EUROCOVER INSURANCE

EUROCOVER MOTORING ASSISTANCE

This service gives protection in the event of vehicle breakdown, accident, fire or theft, or if the only driver is declared medically unfit to drive. Benefits include roadside assistance, replacement vehicle, spare parts dispatch, additional hotel accommodation, vehicle repatriation, legal expenses, and much more.

EUROCOVER PERSONAL TRAVEL INSURANCE

Cover includes insurance against cancellation or curtailment, medical expenses, emergency repatriation, personal accident, theft of personal property and money, travel delay and personal public liability.

The policy can be extended world-wide and there is special winter sports cover.

EUROCOVER PREMIER PROTECTION

This service combines both Motoring Assistance, but with an increased level of cover, and Personal Travel Insurance. It covers up to five persons in a vehicle for a single premium and is the best assistance package available.

■ ESSENTIAL DOCUMENTS, MOTORING ACCESSORIES AND INFORMATION

The RAC is authorised to issue certain official documents which you may need abroad.

Your UK driving licence is recognised in most countries but there are a few exceptions. In these cases we can issue you with an International Driving Permit. If you are driving a UK-registered hired or leased vehicle we will issue you with a Vehicle on Hire Certificate. When you are camping or caravanning we can issue a Camping Card International to cover your party. Some countries stipulate certain accessories as mandatory, such as a red warning triangle, fire extinguisher, first-aid kit or a set of spare bulbs. We can supply all these items, plus headlamp beam deflectors, a motoring emergency kit and even snow chains for hire or purchase.

We can also supply the full range of RAC guides and atlases and remind you of the legal requirements for motoring abroad.

Call free on 0800 765 711 for a copy of the RAC Travel Services brochure or 0800 550 055 to make arrangements.

RAC Publishing

As RAC Publishing, the RAC is a leading publisher of guides and, in conjunction with Bartholomew, the best road atlases for motorists.

The large size floppy *RAC/Bartholomew Road Atlas of Britain 1995* is spiral bound for easy handling and has detailed mapping at 3 miles to 1 inch.

The *RAC/Bartholomew Comprehensive Road Atlas* includes a host of motoring information including 39 town plans, each with a street index, as well as clear mapping at 4 miles to 1 inch. The *RAC/Bartholomew Route Planning Map of Britain* at a scale of 1:550,000 (approximately 9 miles to 1 inch) is an essential planning tool for home or office.

There is an RAC accommodation guide to suit every Member. *RAC Inspected Hotels* is a comprehensive directory to nearly 6,000 hotels, all inspected and approved. Budget travellers and holiday makers will find *RAC Bed & Breakfast* invaluable for booking their accommodation. For open-air enthusiasts, we have *RAC Camping & Caravanning*, or *RAC Farmhouse Accommodation*.

For foreign travel, rely on *Hotels in Europe*, our in-depth coverage of France in *Hotels in France*, or *Camping & Caravanning in Europe*, which covers 3,000 sites in 20 countries.

RAC publications are available in shops throughout the country, or you can order by mail. Please write or phone for a catalogue to RAC Publishing, 39 Milton Park, Abingdon, Oxon OX14 4TD. Tel: (01235) 834885.

RAC Hotel Services

The RAC has been recommending hotels since 1904, and in 1947 the Star Classification Scheme was introduced. This internationally recognised scheme acts as a shorthand to show the level of facilities and services offered. Hotels are classified from 1 to 5 stars and there are over 3,000 RAC Appointed hotels in the British Isles graded using this system. All are inspected annually by our team of hotel inspectors.

If you want a simple and less expensive place to stay at, look for the RAC 'Listed' small hotels and guest houses, of which there are almost 2,000 throughout the UK. Like the Appointed hotels, these are inspected annually.

RAC Insurance Services

■ RAC MOTOR INSURANCE
MEMBERSHIP BENEFITS

As an RAC Member you have access through RAC Insurance Services to a panel of over 20 leading UK insurers. We can offer you the best-value motor policy that will most exactly match your requirements, no matter what car you drive or whatever your personal driving circumstances. Plus, in most cases, there are special advantages just for RAC Members:

- First Response – the RAC's unique 24-hour claims assistance service. From the moment you call us you will have access to your own Claims Assistance Advisor, who will handle your claim from start to finish. In the event that your car cannot be driven, we will make sure it is recovered from the roadside. We will even arrange alternative transport or overnight accommodation and pass on messages to family and friends to tell them you are safe
- double cover for items lost or damaged from an insured vehicle

These are in addition to the extra benefits of:

- premiums payable by credit card or instalments
- our unique Motorists Legal Protection Plan, providing an insured loss recovery and personal injury claims service as well as a 24-hour legal advice helpline
- 24-hour Legal Helpline. Access to our 24-hour Helpline for immediate, confidential advice on any motoring-related legal matter.
- We are open when you need us – for a no-obligation quotation, call our dedicated customer advisors on 0462 435 447 and quote reference 576, Mon-Fri 9am-5pm, Sat 9am-1pm.

■ RAC HOME CONTENTS INSURANCE

RAC Members qualify also for the best value home contents insurance. RAC HomeCare is a home contents insurance which offers you more features and a choice of three levels of cover – all at a highly competitive price.

Premiums are based on the number of bedrooms you have and the district in which you live. In addition to the special RAC Members' discount, in most cases there are also discounts for:

- householders over the age of 55
- fitting certain security precautions
- joining a Neighbourhood Watch Scheme

For a no-obligation quotation call free on 0800 515 505 ext 316, Monday to Friday 8.30am to 7.30pm, Saturday 9am to 1pm.

■ CARAVAN AND SMALL CRAFT INSURANCE

If you have a caravan or small craft, you can also take advantage of a special Members' discount of 10%. Call 0462 421414 for a free, no-obligation quotation.

■ RAC INSURANCE SERVICES NATIONAL REGIONAL CENTRES

As an alternative to calling the numbers above, you may contact your nearest insurance office, open during normal office hours.

Belfast: (01232) 232640
Cardiff: (01222) 610100
Glasgow: 0141-221 5665

RAC Signs Service

If you are organising an event or a show and need signs to direct people to your venue, the RAC Signs Service can help you. We offer a complete service and will put up temporary signs for events of all types. Call the Signs Service free on 0800 234 810 Monday to Friday 9am to 5pm to discuss your requirements.

RAC Privileges and Discounts

As an RAC Member you have access to some very worthwhile special offers and discounts arranged for you by the RAC with major companies and available all over the UK.

■ AUTOGLASS

For a fast, professional glass repair and replacement service, call Autoglass free on 0800 36 36 36. RAC Members are entitled to a substantial discount on all standard stock replacement items.

■ BEAULIEU NATIONAL MOTOR MUSEUM

Here's a fascinating day out in the New Forest for the whole family, and as an RAC Member you can save money on the entrance fee for your party – just show your membership card.

■ BUPA

Medical treatment at a time to suit you, a choice of hospitals and specialist, a private room – these are just some of the advantages of BUPA care. As an RAC Member you can join the RAC-BUPA group and get a discount on the basic rate.

The Disabled Drivers' Association
registered charity number: 254544

The UK organisation for disabled drivers and passengers offers you:

ADVICE - personal and confidential

INFORMATION - publications & 'Magic Carpet'

CAMPAIGNS - working on your behalf

BENEFITS - cost concessions

FRIENDSHIP - local groups

JOIN US TODAY: write, phone or fax us
DDA, Ashwellthorpe, Norwich, NR16 1EX
Tel: 01508 489 449, Fax: 01508 488 173

■ EURODOLLAR

An exclusive new service is now available to RAC Members from Eurodollar, who will provide a car within two hours of your call at preferential rates.

■ HOMETUNE

You can get a discount off any Hometune service under a special deal arranged for RAC Members. Find your nearest Hometune operator in your local telephone directory.

■ NATIONAL TYRES AND AUTOCARE

As an RAC Member you can enjoy a discount from National on any brand of tyre, and on all other products, including exhausts, batteries, brakes, clutches, shock absorbers and radiators. You will also get a discount on an MOT or service.

■ SUPERDRIVE

Superdrive has introduced some excellent deals and discounts for RAC Members on servicing, MOTs, tyres, exhausts, clutches, steering and brake repairs. Look in your Yellow Pages under Exhausts for your nearest centre.

PROTECTING THE INTERESTS OF THE MOTORIST

■ RAC PUBLIC AFFAIRS DIVISION

Since its foundation in 1897 the RAC has worked tirelessly to improve conditions for motorists.

We campaign for new legislation in the UK and EC to improve road traffic law and for better and safer roads. We fight for the rights of motorists and try to ensure that their needs are properly recognised in the environmental debate, and in discussions about how to tackle the growing burden imposed by congestion. A continuing task is to try to keep the overall cost of motoring down and ensure that motorists get value for money.

The RAC's Public Affairs Division exists primarily to pursue these objectives, within the context of the RAC's overall mission, which includes the commitment, 'To champion socially responsible developments in motoring matters, anticipating the needs of all motorists, and promoting an independent voice for them'.

Parking and the Disabled

The Orange Badge Scheme

The Orange Badge Scheme provides a national arrangement of parking concessions for people with severe walking difficulties who travel either as drivers or passengers, registered blind people, and people with very severe upper limb disabilities who regularly drive a vehicle but cannot turn a steering wheel by hand. It allows badge holders to park closer to their destination. The national concessions apply only to on-street parking.

You Can Get a Badge if:

- you receive a mobility allowance
- you receive the higher rate of the mobility component of the Disability Living Allowance
- you are registered blind
- you use a vehicle supplied by a government department
- you get a grant towards your own vehicle
- you receive a War Pensioners' Mobility Supplement
- you have a permanent and substantial disability which means you are unable to walk or have very considerable difficulty in walking. In this case your doctor may be asked to answer a series of questions to help the local authority determine whether you are eligible for a badge. People with a psychological disorder will not normally qualify unless their handicap causes very considerable difficulty in walking
- you have a severe disability in both upper limbs and regularly drive a motor vehicle but cannot turn the steering wheel of a motor vehicle by hand even if that wheel is fitted with a turning knob.

(NB children under 2 years of age do not qualify for a badge because they would not normally be expected to be able to walk independently.)

Where and How to Apply

If you think you may be entitled to a badge you should apply:
- to the Social Services Department – in England and Wales – of your local County, Metropolitan District, or London Borough Council
- to the Chief Executive – in Scotland – of your local Regional or Island Council.

Your local authority will decide if you are eligible for a badge. There is no right of appeal against its decision if you do not meet the eligibility conditions.

The Orange Badge is now designed in the form of a personal passport-type document and has space for a photograph of the holder to be displayed. Your application should, therefore, be accompanied by two photographs, each signed on the back. You may send passport-type photographs taken from self-service booths or any suitable photographs cut down to an appropriate size.

Where the Scheme Does Not Apply

In four central London boroughs: the City of London, the City of Westminster, Kensington and Chelsea, and part of Camden, there are acute parking problems. The four authorities operate their own independent concessionary schemes for people who live or work in their areas. They also offer a limited range of concessions to other disabled people, details of which can be obtained from the respective authorities.

- On the road systems at some airports (for example, Heathrow)
- In certain town centres, where access may be prohibited or limited to vehicles with special permits.

With the exception of the above, the Orange Badge Scheme applies throughout England, Scotland and Wales.

Where to Park

Badge holders may park free of charge and without time limit at on-street parking meters and 'pay-and-display' on-street parking. The Orange Badge must be displayed.

Badge holders may park for as long as they wish where others may park for only a limited time. The Orange Badge must be displayed.

Badge holders may park on single or double yellow lines for up to three hours in England and Wales, and without any time limit in Scotland, except where there is a ban on loading or unloading. The Orange Badge must be displayed, and in England and Wales the special orange parking disc showing the time of arrival must also be displayed.

If in doubt, display the parking disc.

Never park where it would cause an obstruction

or danger to other road users. Your vehicle could be removed by the police, and you could be prosecuted and your badge withdrawn.

The driver must move the vehicle if requested to do so by a police officer or a traffic warden in uniform.

RED ROUTES

Red Routes are main roads in urban areas which are subject to special controls on stopping, loading and unloading. Special parking provision will be made for Orange Badge holders. You should always check with signs to see what concessions are available.

WHERE NOT TO PARK

You must not park:
- when a ban on loading or unloading is in force (indicated by one, two or three yellow marks on the kerb and at times shown on post-mounted plates)
- where there are double white lines in the centre of the road, even if one of the lines is broken
- in a bus lane during its hours of operation
- in a cycle lane during its hours of operation
- on any clearway
- on zebra or pelican crossings
- on zigzag markings before and after zebra or pelican crossings
- in parking places reserved for special users, e.g. residents, taxis, cycles and those using loading bays
- in suspended meter bays or when use of the meter is prohibited.

You should also not park:
- where it would cause a danger to others, e.g. at school entrances or bus stops
- where it would make it difficult for others to see clearly, e.g. close to a junction
- where it would make the road narrow, e.g. by a traffic island or where roadworks are in progress
- where it would hold up traffic, e.g. in narrow stretches of road or blocking vehicle entrances
- where emergency vehicles stop or go in and out, e.g. entrances to hospitals or doctors' surgeries.

HOW TO USE THE BADGE

You must display the new-style passport-type badge on the dashboard or fascia panel of a vehicle with the front facing forward so that the relevant details (e.g. name of badge holder) are legible from outside the vehicle when using the parking benefits. The badge should be removed from the vehicle at all other times. (The old-style badge should continue to be displayed on the nearside of the front windscreen until it expires.)

Similar badges given to institutions caring for disabled people must not be used by able-bodied members for their own benefit and should be removed when the vehicle is not being used for the benefit of disabled people.

Badges last for three years only. When you need a new one apply to the issuing authority for reassessment well before the badge expires.

You must return the badge to the issuing authority if you no longer need it. You should never give the badge to someone else.

HOW TO USE THE SPECIAL PARKING DISC

In England and Wales you will need a parking disc (obtainable from the issuing authority) when you park on yellow lines or in a reserved parking place for badge holders which has a time limit. The disc must be displayed every time you park and set to show the time of arrival. Disabled people living in Scotland who visit England and Wales should be able to get this disc from their local Regional or Island Council.

GIVING A VEHICLE REGISTRATION NUMBER

A badge is issued to an individual for the vehicle or vehicles which he drives or in which he is carried as a passenger.

The issuing authority must be given the registration number of the vehicle or vehicles (or, in the absence of a registration number, sufficient information to enable identification of the vehicle to be made) when a badge is applied for.

YOUR DUTIES AS A BADGE HOLDER

The purpose of the Scheme is to allow you to visit shops and other places. It is your responsibility to ensure that the badge is used properly.

You must not let other people use the badge. To reduce the risk of this happening accidentally, you should remove the badge whenever you are not using the parking concessions.

The details on the front of the badge (e.g. holder's name and the date of expiry) have to remain legible. If they become unreadable, the badge must be returned to the local authority for re-issue.

You should not take advantage of the concessions if you do not intend to leave the car – use a car park instead. Don't use the badge simply to allow able-bodied people to take advantage of the benefits while you sit in the car.

If you mis-use, or allow others to mis-use, your badge, it can be withdrawn. It is a criminal offence, liable to a hefty fine, for able-bodied people to use a badge. It is also illegal to drive a vehicle displaying an Orange Badge unless the badge is properly issued and displayed.

If you are a disabled driver and your disability is such that it is likely, or may become likely, to affect your ability to drive (even if your car is adapted) the law requires you to inform the Driver and Vehicle Licensing Agency, Swansea SA99 1TU.

Othe Benefits for Badge Holders

Vehicles cannot be wheel clamped on the public highway if a current badge is displayed on the vehicle.

Badge holders may be allowed access to some town centres where vehicle entry is restricted.

In many areas local authorities provide reserved parking places for badge holders. You should use these spaces in preference to parking on yellow lines. You must always display a valid badge when occupying one of these spaces, and if a time limit is in force a parking disc must also be displayed.

Some local authorities also waive charges in their off-street car parks. Please check with notices in the car park or with an attendant.

Badge holders are exempted from certain tolls (see 'Toll concessions' on page 40).

Some other European countries allow disabled visitors to take advantage of the parking concessions provided for their own citizens by displaying the Orange Badge (see 'Reciprocal parking arrangements' on page 54).

Motability

Gate House, West Gate Harlow, Essex CM20 1HR
☎ 01279-635666 (customer services/helpline)

Motability is a voluntary organisation and registered charity which was set up in 1977 to see that recipients of the higher-rate mobility component of the Disability Living Allowance (DLA) or War Pensioners' Mobility Supplement (WPMS), and owners of DSS three- and four-wheelers, get the maximum value for money when purchasing a car or wheelchair.

The organisation has arranged special terms and discounts with car and wheelchair manufacturers on a variety of models, and a finance company has been set up to offer favourable terms. Hire purchase facilities are available to help people to buy new and used cars, and new wheelchairs.

Contract Hire

Under this scheme, disabled people, including children from the age of five, can get a new vehicle from a wide range of approved manufacturers. All maintenance and servicing costs are covered, together with comprehensive insurance and recovery service membership.

There is also a 'loss of use' insurance scheme, which provides a weekly payment, equivalent to the Disability Living Allowance, if the car is off the road for more than a week. Customers agree to pay over all of their allowance for the duration of the three-year hire agreement. For cars at the lower end of the price scale, available with no advance payment, this is the only financial commitment the disabled person has to make, apart from paying for petrol and oil. For other models, the customer may be required to pay a deposit, because the allowance alone is insufficient to cover the cost of providing the vehicle for the three-year term. The amount of the advance payment will vary according to the cost of the vehicle.

Hire Purchase

For people who want to buy their cars, Motability offers a hire purchase scheme, under which finance can be provided towards new and used cars in return for all or part of a person's allowance. To obtain a new car through this scheme, customers must receive the Disability Living Allowance for a minimum of four years from the date of the contract. Number plates, delivery charges, maintenance, repairs and fully comprehensive insurance are all extra costs that have to be paid under the Hire Purchase scheme. The finance provided towards the car is offered at favourable rates of interest.

To buy a used car, customers must have at least two years' DLA, and the car must be under four years old, have done less than 40,000 miles (30,000 if an Alfa Romeo, Lada or Seat), and pass an inspection. Certain vehicles cannot be purchased through this scheme – check with Motability. The dealer from whom the car is bought must either hold a recognised manufacturer's franchise (i.e. be an officially recognised dealership) or be a member of the Retail Motor Industry Federation or of the Scottish Motor Traders Association.

Cars supplied on both the Contract Hire and Hire Purchase schemes can be adapted where necessary to suit the needs of the disabled person, although the Hire Purchase scheme is more suitable for very heavily adapted vehicles. If required, Motability can advise on driving assessment organisations and adaptation needs.

If a disabled person cannot meet the full cost of putting a car on the road, including any adaptations and the cost of driving lessons, Motability may be able to provide financial help.

For more information on the choices available and the costs involved, write to or telephone the Customer Services Department on the number given above.

INDEPENDENT DRIVING ASSESSMENT CENTRES

BANSTEAD MOBILITY CENTRE

Damson Way, Orchard Hill, Queen Mary's Avenue, Carshalton, Surrey SM5 4NR
☎ 0181-770 1151

The Banstead Mobility Centre is associated with the Queen Elizabeth's Foundation for Disabled People. Clients are welcome to refer themselves to the centre, and while an appointment must be made before a visit, anyone may telephone or write for general information. Residential accommodation is available on the site.

An application form will be sent to all prospective clients for wheelchair, passenger or driving assessments, and on receipt of this the applicant's GP or consultant will be contacted for background medical information. (A medical form is required for driving assessments.)

The centre has a Ford Transit with joystick controls for driving from a wheelchair, and a range of modified cars, pavement vehicles and wheelchairs. It features an open-air tarmac road system with crossings, junctions and varying kerb heights.

Types of assessment

1) A very full driving assessment (£90), taking up to six hours, is designed for those affected, for example, by a stroke, spina bifida, cerebral palsy or a head injury. The client is seen by a medical consultant, orthoptist, psychologist, therapist and driving instructor. The assessment includes a track test in a suitably modified car, and experienced drivers also have a road test, if appropriate. A written report is sent to the client and to the doctor.

2) The adaptation assessment (£50) takes two hours and is designed for those affected, for example, by arthritis, spinal lesions and dystrophies. It helps those wishing to modify a standard vehicle to compensate for their disability. A fleet of ten cars is available for test driving on a private track. Problems of access, seating and storage of wheelchairs are investigated by a therapist. A driving instructor assesses each client in a static unit equipped to measure steering strength, brake pressures and reaction times. In-car testing is an important part of the assessment. A report is sent to the client and to the doctor.

3) Assessment of lightweight manual wheelchairs (£30).

4) Advice is provided on all aspects of driving and also on wheelchairs and pavement vehicles. Training sessions include transferring to/from a wheelchair (£12 per hour).

5) Driving lessons are available on specially equipped vehicles (£14 per hour). Five-day driving courses can be arranged. Residential driving holidays are organised for learner drivers with a disability.

DISABILITY ACTION

2 Annadale Avenue, Belfast BT7 3UR,
☎ 01232-491011

A professional assessment service is provided for any individual wishing to drive. Recommendations for suitable models of cars, controls and adaptations can be made, using an in-car electronic assessment unit to test reaction times, physical strength, eyesight and other relevant factors. There are also driving school facilities, which include three hand-control-operated automatic cars for driving tuition.

Driver assessment is also available outside of Belfast, as the assessment unit is mobile. One of the cars available for driving tuition is based in Londonderry.

MOBILITY ADVICE AND VEHICLE INFORMATION SERVICE (MAVIS)

Transport Research Laboratory, Crowthorne, Berkshire RG11 6AU, ☎ 01344-770456

The Mobility Advice and Vehicle Information Service was set up by the Department of Transport to provide practical advice on driving, car adaptations and car choice for people with disabilities and for elderly people, both as drivers and passengers. The centre offers:

- a full driving ability assessment, including advice on car adaptations and the opportunity to test drive a range of vehicles
- consultation and advice on car adaptation and a test drive
- consultation and advice on car adaptations for passengers.

The centre has facilities to measure a person's strength, steering force, reaction times, grip, correct seating position and other important factors. A written report is sent out after the assessment.

You can inspect and/or test drive vehicles with a variety of adaptations, equipment and accessories. The models are regularly updated and include most currently available vehicles.

The equipment includes a range of hand controls for manual and automatic transmission cars, power-assisted steering, joystick steering, vacuum brakes, left-foot accelerators, handbrake/gear selector modifications and special mirrors. There is also a range of car access aids, such as swivel seats, and wheelchair stowage equipment.

Test driving, which takes place under expert supervision in vehicles fitted with dual controls, is carried out on the private road system at the centre and may include a session on public roads. It is important to have the right type of assessment.

A free information service is available on all aspects of transport (public and private) and outdoor mobility for people with disabilities and for elderly people.

MOBILITY INFORMATION SERVICE (MIS)
Unit 2A, Atcham Estate, Shrewsbury, Shropshire SY4 4UG, ☎ 01743-761889

This specialist service gives guidance and advice on aspects of mobility, including wheelchairs, choice of vehicles, mobility allowance, parking privileges and financial/tax concessions.

The MIS offers a driver assessment service, with a static simulator unit which can test reaction time and physical strength. The simulator is equipped with a range of adaptations and hand controls, is able to test braking pressure and steering torque, and provides a computerised printout result. A private track allows individuals to try an adapted vehicle (two cars are available) as part of the assessment. A written report is provided.

MIS publications include 'Adaptations for the disabled driver' (£1.25), 'Wheels under you' (£1), and information sheets and road test reports on suitable cars for disabled drivers, available as an information pack (£3). MIS also provides a 'New Driver's Pack' (£5), which is tailored to the requirements of each person (send an s.a.e. for details).

WALES DISABLED DRIVING ASSESSMENT CENTRE
18 Plas Newydd, Whitchurch, Cardiff CF4 1NR
☎ 01222-615276

The centre has a static simulator fitted with all the usual controls, both hand and foot. It is capable of measuring strength of grip, steering force and braking capabilities, enabling advice to be given on the type of vehicle and the controls needed. Ease of vehicle entry is discussed, and swivel and slide seats can be seen and considered. Where required, driver-vision screening tests are used, which can prove of great value to the safety of the driver.

Professional advice is offered on the type of vehicle and the suitability of the various controls that are available. Information is provided on all aspects of outdoor mobility, benefits, parking permits, driving instructors and insurance, and on fitters who can complete adaptations required. The centre has an automatic car with power-assisted steering, fitted with hand controls and steering wheel spinner for in-car assessment with an approved driving instructor.

■ REGIONAL DRIVING ASSESSMENT CENTRES

CORNWALL
CORNWALL FRIENDS MOBILITY CENTRE
Tehidy House, Royal Cornwall Hospital, Truro TR1 3LJ   ☎ 01872-260060

DERBYSHIRE
DERBY DISABLED DRIVING CENTRE
Kingsway Hospital, Kingsway, Derby DE3 3LZ
☎ 01332-371929

DEVON
DEVON DRIVERS' CENTRE
G A House, Westpoint, Clyst St Mary, Exeter EX5 1DJ   ☎ 01392-444773

HERTFORDSHIRE
HERTFORDSHIRE ASSOCIATION FOR THE DISABLED, DRIVING ASSESSMENT AND INSTRUCTION CENTRE
The Woodside Centre, The Commons, Welwyn Garden City AL7 4DD   ☎ 01707-324581

KENT
CAR MOBILITY ADVICE CENTRE
Home Equipment Loan Point, Preston Hall, Aylesford ME20 7NJ   ☎ 01622-710161 ext 2241

LANCASHIRE
DISABLED DRIVERS' VOLUNTARY ADVISORY SERVICE
18 The Roods, Warton, nr Carnforth LA5 9QG
☎ 01524-734195/823189

MERSEYSIDE
CLATTERBRIDGE DRIVING ASSESSMENT SERVICE
Wirral Limb Centre, Clatterbridge Hospital, Bebington, Wirral L63 4JY   ☎ 0151-604 7439

NORFOLK
KILVERSTONE MOBILITY CENTRE
Kilverstone, Thetford IP24 2RL   ☎ 01842-753029

OXFORDSHIRE
MARY MARLBOROUGH CENTRE
Windmill Road, Headington, Oxford OX3 7LD
☎ 01865-227600

STAFFORDSHIRE
MID-STAFFORDSHIRE DRIVING ASSESSMENT SERVICE
Cannock Chase Hospital, Brunswick Road, Cannock WS11 2XY   ☎ 01543-576424

TYNE AND WEAR
HUNTER MOOR REGIONAL REHABILITATION CENTRE
The Mobility Centre, Hunters Road, Newcastle upon Tyne NE2 4NR   ☎ 0191-221 0454

WEST MIDLANDS
MOBILITY SERVICE
Mobility Service Co-ordinator, Hillcrest Rehabilitation Centre, Moseley Hall Hospital, Moseley, Birmingham B13 8JD,
☎ 0121-442 4321 ext 306,

YORKSHIRE
BARNSLEY REGIONAL MOBILITY CENTRE
Moreland Avenue, Barnsley, South Yorkshire
S70 6PH ☎ 01226 201101/284710/245873

WALES
ROOKWOOD DRIVING ASSESSMENT CENTRE
Rookwood Hospital, Fairwater Road, Llandaff, Cardiff CF5 2YN ☎ 01222-566381 ext 3776 (morning), ext 3755 (afternoon)

WALES DISABLED DRIVERS' ASSESSMENT CENTRE, NORTH WALES
Ysgol-Gogarth, Nant-y-Gamar Road, Llandudno
LL30 1YF ☎ 01492-872036

SCOTLAND
EDINBURGH DRIVING ASSESSMENT CENTRE
Astley Ainslie Hospital, 133 Grange Loan, Edinburgh EH9 2HL ☎ 0131-537 9193

CAR MANUFACTURERS OFFERING DISCOUNTS AND CONCESSIONS FOR DISABLED DRIVERS

The following car manufacturers offer some kind of discount on new cars for disabled drivers. The majority work through the Motability Contract Hire and Hire Purchase schemes as described in the previous section of this guide. Some manufacturers, however, also have schemes which are open to anyone who is registered disabled, and not just those who receive the higher-rate mobility component of the Disability Living Allowance or War Pensioners' Mobility Supplement. There are also some manufacturers that will offer a preferential discount for disabled people who want to buy a new car outright.

Where specific figures have been given for discounts, these were correct at the time of writing but are liable to change at any time, so they should be seen only as a guide. This is a constantly changing scene, and manufacturers not mentioned here may, in the future, join the Motability system.

■ BMW
BMW (GB) Ltd, Ellesfield Avenue, Bracknell RG12 8TA ☎ 01344-426565

Discounts are available through individual dealers. For more information contact your local dealer.

■ CITROEN
Leasing and Rental Manager, Citroën UK Ltd 221 Bath Road, Slough SL1 4BA ☎ 01753-822100

Virtually the full dealer discount is available on the whole Citroën range for registered disabled people who want to buy outright or on hire purchase.

Discounts are available also through the Motability Hire Purchase scheme, and Citroën is working to get full status on the Motability Contract Hire scheme for 1995.

■ DAIHATSU
Daihatsu (UK) Ltd, Poulton Close, Dover CT17 0HP ☎ 01304-213030

The Applause and Charade ranges (except for the Charade GSXi) are offered at a 10% discount through the Motability Hire Purchase scheme. Other models are available on hire purchase on a supply and demand basis, though discounts have to be negotiated with local dealers.

■ FIAT
Fiat Mobility Administration Office, PO Box 62, Dunstable LU5 5JX ☎ 01582-863634

A 9% discount and a reduced delivery charge are offered on all new cars supplied by either the Contract Hire or Hire Purchase Motability schemes.

Customers may also purchase a new car through Fiat's own Outright Purchase Scheme. The buyer must still be in receipt of the higher-rate mobility allowance, but the duration of the award is not important. This offer enjoys the same 9% discount but the dealer will charge the full amount for delivery and plates.

■ FORD
Ford Motor Company Ltd, Ford Information Services, Wonderman Kato Johnson, Unit 3, Genesis Business Park, Albert Drive, Woking, Surrey, GU21 5RW ☎ 0800 111 222

The majority of Ford models are available for disabled people, or for their parents, spouses or carers, at a discount through the Motability Contract Hire and Hire Purchase schemes. In addition, any registered disabled person may purchase a car at a substantial discount through the Ford Motability Outright Purchase Scheme. For further details contact your nearest Ford main dealer, or call the Freephone number given above.

■ HYUNDAI
Hyundai Car (UK) Ltd, St John's Place, Easton Street, High Wycombe HP11 1NL ☎ 01494-428600

All models are available at discounted prices

through the Motability Hire Purchase scheme, and all except the Sonata through the Contract Hire scheme. The entire range can be adapted, and the majority of Hyundai dealers are Motability-approved.

■ KIA

Kia Cars (UK) Ltd, 77 Mount Ephraim, Tunbridge Wells TN4 8BS ☎ 01892-513454

Discounts of 2.5% to 10% on the Kia Pride and up to 18% on the Kia Mentor are offered through the Motability Contract Hire and Hire Purchase schemes.

■ LADA

Lada Cars, 3120 Park Square, Birmingham Business Park, Birmingham B37 7YN ☎ 0121-717 9000/0345 44 66 44 (information hotline)

Samara and Niva vehicles are available at discounted prices through the Motability Hire Purchase scheme. Contact your local dealer for more information.

■ NISSAN

Motability Administration Department, Nissan Motor GB, The Rivers Office Park, Denham Way, Maple Cross, Rickmansworth WD3 2YS ☎ 01923-899466

Nissan is able to offer any registered disabled person various discounts ranging between 5% and 8% on its cars and commercial vehicles. Models are available at a discount through the Motability Contract Hire and Hire Purchase schemes.

■ PEUGEOT

Peugeot Motability Departments, Peugeot Talbot Motor Company PLC, Aldermoor House, PO Box 227, Aldermoor Lane, Coventry CV3 1LT ☎ 01203-884000

Peugeot provides four alternative methods of acquiring the ownership or use of a new car: the Motability Hire Purchase and Contract Hire schemes, conventional hire purchase and outright purchase. Discounts are available throughout the Peugeot range, the amount depending on which model you choose. Accessories and special adaptations are not subject to a discount.

Whether you are in receipt of a mobility allowance or not, if you are registered disabled you can arrange your own hire purchase or pay cash and still qualify for a Peugeot discount. You should consult your Peugeot dealer for full details of the various deals available.

■ PROTON

Proton Cars (UK) Ltd, Proton House, Royal Portbury Dock, Bristol BS20 0NH ☎ 0275-375475

A discount of 8% is given to all Motability customers through the Hire Purchase and Contract Hire schemes. Car adaptation can be arranged through Proton dealers. Proton also offers a 'Motability Special', a model available at a competitive price and which has power-assisted steering and automatic transmission.

■ RENAULT

Renault UK Ltd, c/o PO Box 527, Houghton Regis, Bedfordshire LU5 5UR ☎ 0800 387626

The Renault Disabled Motorist Programme is open to all registered disabled people and offers up to a 10% discount, depending on the model chosen, from a range including the Clio, 19, Savanna Estate, Espace and Laguna. There is also a wide range of vehicles available on the Motability Contract Hire scheme with competitive initial rentals.

■ ROVER

Rover Motability, PO Box 368, Dunstable LU5 5YR ☎ 0345 045310

Discounts are available on most Rover models through the Motability Hire Purchase and Contract Hire schemes. If you do not qualify for the higher-rate mobility component of the DLA or the WPMS, or you would rather not surrender your allowance or supplement, Rover can still help. Providing you are a registered disabled person, you may qualify for a discount through the Rover Mobility Scheme.

Discounts are also available to relatives of registered disabled people, as long as the car is to be used primarily for the disabled person's personal transport.

■ SAAB

SAAB GB, SAAB House, Globe Park, Marlow, Buckinghamshire SL7 1LY ☎ 01628-486977

Discounts are available for holders of mobility allowance certificates. All discounts are arranged through SAAB dealers.

■ SEAT

Seat UK Ltd, Seat House, Gatwick Road, Crawley, West Sussex RH10 2AX ☎ 01293-514141

Discounts are available on most Seat models through the Motability Hire Purchase scheme.

■ SKODA

Skoda Automobile UK Ltd, Garamond Drive, Great Monks Street, Wymbush, Milton Keynes, Buckinghamshire MK8 8NZ ☎ 01908-264000

Discounts are available on models in the Favorit

range through the Motability Hire Purchase and Contract Hire schemes. The same discounts apply if you want to buy a car outright.

■ SUBARU
Subaru (UK) Ltd, Ryder Street, West Bromwich, Birmingham B70 0FJ ☎ 0121-522 2000
A 2.5% discount is available off the retail cost of a Subaru car for any registered disabled person, with proof of registration.

■ SUZUKI
Sales Department, Suzuki GB PLC, 46/62 Gatwick Road, Crawley, West Sussex RH10 2XF ☎ 01293-518000
As we went to press, Suzuki were planning to offer discounts through Motability from early 1995.

■ TOYOTA
Toyota GB Ltd, The Quadrangle, Redhill, Surrey RH1 1PX ☎ 01737-768585
A discount is available on most models for recipients of the higher-rate component of the DLA for cash purchases or through the Motability Contract Hire and Hire Purchase schemes.

■ VAUXHALL
Vauxhall Motability Programme, Carlton House, 62/64 High Street, Houghton Regis, Dunstable LU5 5BJ ☎ 01582-861888
People in receipt of the higher-rate mobility component of the DLA or the WPMS are offered special rates by Vauxhall. Preferential prices or low-cost contract hire are available on many Corsa, Astra, Cavalier and Omega models through your local Vauxhall Motability Registered Dealer. In addition, Orange Badge holders also qualify for preferential prices through Vauxhall's own Mobility Plan.
Every Vauxhall Motability Registered Dealer has a Motability consultant who can give advice and guidance on the financial options and the car most appropriate to your needs, as well as organise any conversions or adaptations that are required.

■ VOLKSWAGEN
V.A.G. (UK) Ltd, Yeomans Drive, Blakelands, Milton Keynes, Buckinghamshire MK14 5AN ☎ 0800 333666
Discounts are available on models in the Polo, Golf, Vento, Passat and Caravelle ranges through the Motability Hire Purchase and Contract Hire schemes. The same discounts apply if you want to purchase a car outright or use the Volkswagen Solutions scheme, through which, for agreed monthly charges, you can encompass the costs of finance, service, maintenance and tyres.

■ VOLVO
Volvo Car UK Ltd, Globe Park, Marlow, Buckinghamshire SL7 1YQ, ☎ 01628-477977
There is a 7.5% discount from Volvo dealers on all models for disabled people through the Motability Contract Hire and Hire Purchase schemes.
Those people who do not wish to use the Motability schemes but are in receipt of the higher-rate mobility component of the DLA may obtain a minimum discount of 7.5% on all Volvo 400s (except diesel) and 7.5% on all Volvo 800/900s and 400 diesel.

FIRMS THAT MANUFACTURE OR CONVERT VEHICLES

ADAPTACAR
5 Cooks Cross, South Molton, Devon EX34 4AW ☎ 01769-572785
Produces a hand-operated or electric hydraulic hoist to lift wheelchairs into the rear of a hatchback/estate. Also produces brake and steering aids.

AUTOMOBILE AND INDUSTRIAL DEVELOPMENTS LTD
Queensdale Works, Queensthorpe Road, Sydenham, London SE26 4PJ ☎ 0181-778 7055
Produces hand controls for automatic cars.

ALFRED BEKKER
Kelleythorpe, Driffield, Humberside YO25 9DJ, ☎ 01377-241700
Produces hand-control equipment for any make of car, both manual and automatic.

BRIG-AYD CONTROLS
Warrengate Farm, Warrengate, Tewin, Welwyn, Hertfordshire AL6 0JD ☎ 01438-714206
Offers a wide range of conversions, including steering, brakes, clutch and seats. Converts most makes of car.

BRUNELL FORD
Victoria House, Temple Gate, Bristol BS1 6PR, ☎ 0117 - 929 4222
Carries out hand control conversions to any vehicle.

AA CLARK FOR VAUXHALL
195 Clarence Road, Windsor, Berks SL4 5AN, ☎ 0753-863456 Fax: 0753-830103

BROTHERWOOD AUTOMOBILITY LTD
Pillar Box Lane, Beer Hackett, Sherborne, Dorset
DT3 6QP, ☎ 01935-872603
Conversions of Nissan Prairie and Ford Courier. The company also manufactures the Transfer Beam, which gives the front passenger seat of a car a rotating facility within the car and extends the seat over the sill of the car door. This can be fitted to Nissan Prairie, Nissan Micra, Ford Fiesta and Ford Escort.

CITY MOTOR SERVICES
1 Shakespeare Street, Roath, Cardiff CF2 3DQ
☎ 01222-482804 Fax: 01222-482804
Car conversions: hand controls, electronic seats, power steering, infra red (either 4 or 9 options on steering wheels).

CLARK AND PARTNERS LTD
80/136 Edmund Road, Sheffield S2 4EE
☎ 0114 - 275 5000
Produces a range of hand controls that can be fitted to most vehicles.

CLARKS INDEPENDENT CENTRE
71 New Road Side, Horsforth, Leeds LS18 4JX,
☎ 0113 258 8888
Supplies and fits hand controls, power steering, rotating seats, infra-red switching units and electric clutch systems, driver transferring systems and wheelchair lifts. Full van conversions are available.

CONSTABLES LTD
Mountney Business Park, Westham, Pevensey, East Sussex BN24 5NJ, Tel: 01323-767574 Fax: 01323-767603
Manufactures car wheelchairs that are specifically designed to be electronically lifted into either passenger, driver or back seats. They convert the car so that it is able to do this. Another service they provide is general minibus conversions to specific requirements.

CONVERSIONS LTD
75 Alverston Road, Milton, Portsmouth
PO4 8TG, ☎ 01705-756265
Converts Renault Extra, Traffic and Master ranges.

COWAL (MOBILITY AIDS) LTD
32 New Pond Road, Holmer Green, High Wycombe, Buckinghamshire HP15 6SU,
☎ 01494-714400
Produces the Wymo Wheelchair Hoist and GZ-91 roof rack for automatic wheelchair storage. Also specialises in conversions to a range of vehicles and supplies hand control kits.

DESIGN AND INNOVATION
Unit 55, Alston Drive, Bradwell Abbey, Milton Keynes, Buckinghamshire MK13 9HB,
☎ 01908-226688
Specialises in converting and adapting any vehicle, using its own products.

DEVON CONVERSIONS (CP) LTD
Water Lane, Exeter, Devon EX2 8BY,
☎ 01392-211611
Conversions of Fiat and Volkswagen vans.

JIM DORAN HAND CONTROLS
Unit 2, Parbrook Close, Torrington Avenue, Tile Hill, Coventry CV4 9XB Tel: 01203-460833

FEENY AND JOHNSON (COMPONENTS) LTD
(see also **STEERING DEVELOPMENTS LTD**)
Unit 5, Eastman Way, Hemel Hempstead, Hertfordshire HP2 7HF, ☎ 01442-212918
Provides a variety of conversions, including change-over foot accelerator for automatic cars, re-siting of the parking brake and the fitting of steering-wheel knobs.

HELSDEN & CO (HOLT) LTD (VAUXHALL)
37, Cromer Road, Holt, Norfolk ☎ 0263-713207
Fax: 0263-713207

D.G. HODGE AND SON LTD
Feathers Lane, Hythe End, Wraysbury, Staines, Middlesex TW19 5AN, ☎ 01784-483580
Produces a system to help a disabled driver gain access to the driver's seat from the passenger side while at the same time providing stowage for the wheelchair. Also provides a wide range of driving controls.

GOWRINGS MOBILITY INTERNATIONAL
The Old Barn, 18/21 Church Gate, Thatcham, Berkshire RG13 4PH, ☎ 01635-871502/0800 220878
Produces the Chairman range of vehicles: Vista, Hi-jet, Vanette, Extra, Fiesta, Escort Popular, Escort Elite, Serena and Esprit. These are based on existing production cars and are designed to carry someone travelling in their wheelchair plus three or four other passengers. Some models can accommodate two wheelchair users.

IJK MOBILITY
Guestling, The Street, Bradwell, Braintree, Essex CM7 8EG ☎ 01376-562152

INTERBILITY LTD
Unit 21, Gunnels Wood Park, Stevenage, Hertfordshire SG1 2BH ☎ 01438-747448
Conversions of light vans such as Ford Transit, Volkswagen Transporter and Renault Trafic. Also produces wheelchair lifts and ramps.

THE NEW VOLKSWAGEN CARAVELLE
BY INVATRAVEL CONVERSIONS

Various options of wheelchair access on the super new model, including underfloor electrohydraulic lifts, lowering rear suspension with spring assisted alloy ramp etc.
Full conversions for disabled driver independence.
Prices start from £1475.00 excluding taxes.
Contact Paul Cassidy for brochures or a demonstration.
Invatravel are recognised converters for Volkswagen.

Invatravel Conversions
BASED ON THE VOLKSWAGEN CARAVELLE
66, Knob Hall Lane, Southport PR9 9QS
Telephone: 01704 506608 Fax: 01704 506607

APPEAL

WITH THE RENAULT
DRIVING SCHOOLS PACK OFFER*

FREE £100 CASHBACK

+

**PERSONALISED HEADER BOARD
SIDE DECALS
INSTRUCTORS MIRROR
DUAL CONTROLS**

OR

£350 CASHBACK

* offer subject to change

01792 701801
FENDROD WAY ENTERPRISE PARK   LLANSAMLET SWANSEA

CARCHAIR IN ACTION

Constables understand the needs of the wheelchair user...which is not surprising when you consider our pedigree... with over twelve years' first-hand experience of the mobility market.

The Carchair is a wheelchair system which becomes the front passenger or driver's seat of a standard production motor car. Naturally, this eliminates the need to transfer from wheelchair to car seat and avoids the purchase of a 'special' vehicle.

As a complete unit the Carchair system comprises a specially designed wheelchair and easy-to-operate lift mechanism which can be fitted into a range of vehicles. The wheelchair itself is as much at home around the house as it is in the family car or out-of-doors. Quite simply, no other wheelchair is so adaptable.

Constables LTD

CONSTABLES LTD • DEPT RAC • FREEPOST (BR1032) • WESTHAM • PEVENSEY • EAST SUSSEX • BN24 5BR TELEPHONE: (01323) 767574

FIRMS THAT MANUFACTURE OR CONVERT VEHICLES

INVATRAVEL
66 Knob Hall Lane, Southport PR9 9QS,
☎ 01704-506608
Van and minibus conversions.

KC MOBILITY SERVICES
KC House, Carlinghow 2, Bradford Road, Batley,
West Yorkshire WF17 8LL   ☎ 01924-442386

LYNX HAND CONTROLS LTD
Mansion House, St Helens Road, Ormskirk,
Lancashire L39 4QJ   ☎ 01695-573816
Produces hand controls that are transferable from car to car and fit most automatic cars.

MACKAYS OF DINGWALL (VAUXHALL)
Strathpeffer Road, Dingwall IV15 9QF
Contact: Mike Sim   ☎ 01349-862011
Fax: 01349 -862088
Vauxhall motability specialists. Body repairs,
Vauxhall retail and servicing to a very high standard.

MELLOR COACHCRAFT
Coachbuilding and Special Vehicles, Miall Street,
Rochdale OL11 1HY   ☎ 01706-860610
Conversions of most manufacturers' base vehicles,
including vans produced by Iveco Ford, Mercedes-Benz,
Leyland Daf, Renault, Ford and Volkswagen.

THE MOBILITY WORKSHOP LTD
Unit 12, Barton Business Park, Cawdor Street,
Eccles, Manchester M30 0QR   ☎ 0161-707 9860
Produces hand controls for automatic and manual
vehicles, also conversions such as wheelchair ramps.

NEILL AND BENNETT LTD
7 Wyngate Road, Cheadle Hulme, Cheshire
SK8 6ER   ☎ 0161-485 3149
Conversions of vehicles such as Renault Trafic, Ford
Transit, Peugeot Boxer and Seat Terra. Also produces
wheelchair ramps.

BRIAN PAGE CONTROLS
5 Eversley Way, Thorpe Industrial Estate, Egham,
Surrey TW20 8RG   ☎ 01784-435850/430423
The southern distributor for Alfred Bekker hand
controls. Other minor adaptations are carried out on
request, including an infra-red switching device for
drivers with the use of only one arm.

PALADON LTD
Unit 18, Central City Industrial Estate, Red Lane,
Coventry CV6 5RY   ☎ 01203-638588
Produces a range of cruise controls.

PILCHER-GREENE LTD
Specialist Vehicle Builders, Consort Way, Burgess
Hill, West Sussex RH15 9NA   ☎ 01444-235707

Conversions of most manufacturers' base vehicles.

PINCKNEY MOBILITY
Alma Works, The Street, Takeley, nr Bishops
Stortford, Hertfordshire CM22 6QU
☎ 01279-870834

POYNTING CONVERSIONS
Faraday Road, Churchfields Industrial Estate,
Salisbury, SP2 7NR   ☎ 01722-336048
Conversions of cars and small vans such as Rover
Metro and Maestro, Renault Extra and Seat Vista.

RATCLIFF TAIL LIFTS LTD
Bessemer Road, Welwyn Garden City,
Hertfordshire AL7 1ET   ☎ 01707-325571.
Produces wheelchair lifts.

RESELCO ENGINEERING LTD
Kew Bridge Pumping Station, Green Dragon
Lane, Brentford, Middlesex TW8 0EN
☎ 0181-847 4500/4509
Provides a variety of special controls for most makes of
manual or automatic car.

ROSS AND BONNYMAN (EADIECARE) LTD
Roberts Street, Forfar, Angus, Tayside DD8 3DG
☎ 01307-466262
Produces wheelchair lifts.

ROSS CARE CENTRE
2/3 Westfield Road, Wallasey, Merseyside
L44 7HX   ☎ 0151-653 6000
Supplies and fits a variety of car adaptations, such as
hand controls, foot pedal controls, steering braces, manual
and electronic switches, swivel seats, hoists and ramps.

MR B. STARRITT
111 Killyleagh Road, Killinchy, Newtownards,
Co Down BT23 6TR   ☎ 01238-541550
Fits hand controls.

SVO (SPECIALISED VEHICLE OPTIONS)
Lottage Road, Aldbourne, Marlborough,
Wiltshire SN8 2EB   ☎ 01672-40001

SIMPLEX DRIVING SYSTEMS
Brunswick Business Park, 18 Summers Road,
Liverpool L3 4BL   ☎ 0151-707 1146
Agents for Guidosimplex, a range of hand controls.

SPEKEHALL(VAUXHALL)
Spekehall Road, Speke, Merseyside L24 9HD
☎ 0151 - 486 3668

STARTINS OF BIRMINGHAM
71 Aston Hall Road, Birmingham, B67 JSX
☎ 0121 - 328 0833 Fax: 0121 - 327 4699

Vauxhall motability specialists, sales service, body repairs, master hire leasing, Vauxhall rental.

STEERING DEVELOPMENTS LTD
Unit 5, Eastman Way, Hemel Hempstead, Hertfordshire HP2 7HF ☎ 01442-212918
Specialises in converting manual steering cars to power-assisted steering for mainly small-engined vehicles, both automatic and manual transmission.

SUNGIFT MARKETING SERVICES
Kempston, Bedfordshire MK42 7AF ☎ 01234-841311 Fax: 01234-841328
Manufacture 10 different vehicles for the disabled, which include powered chairs, pavement vehicles and road vehicles.

SWS MOTOR BODIES
Unit 1, Hartford House, Newport Road, Weston Street, Bolton, Lancashire BL3 2AX
☎ 01204-395660
Conversions of Fiat Fiorino van and the majority of manufacturers' base vehicles. Also fits cars with hand controls and undertakes a range of other modifications.

UNIVERSAL MOBILITY LTD
Jordans, Partridge Lane, Rusper, West Sussex
☎ 01293-87019

VAN DESIGN SYSTEMS
Fenrod Way, Swansea Enterprise Park, Swansea
☎ 0792-701801 Fax: 0792-795661

YORKSHIRE MOBILITY SERVICES
Willow Street, Hopwood Lane, Halifax, West Yorkshire HX1 4DH ☎ 01422-352965
Supplies and fits a range of hand controls.

CITY MOTOR SERVICES (CARDIFF)

CONVERSION FROM HAND CONTROLS TO ELECTRIC CLUTCHES & POWER STEERING

MOTABILITY & WALES ASSESSMENT CENTRE APPROVED

1 Shakespeare Street,
(Off City Rd)
Cardiff CF2 3DQ
Tel: (01222) 482804

otability

on the road to freedom

M *obility*
A *dvice and*
V *ehicle*
I *nformation*
S *ervice*

For more information please write to:

MAVIS

Transport Research Laboratory
Crowthorne, Berkshire RG11 6AU
or telephone: (01344) 770456

WHO ARE WE?

The Mobility Advice and Vehicle Information Service was set up by the Department of Transport to provide practical advice on driving, car adaptations and car choice, both for disabled drivers and passengers.

WHAT CAN WE OFFER?

- Assessment and advice on driving ability and car adaptation.
- Consultation and advice on car adaptations.
- The opportunity to try out one or more of the wide range of adapted vehicles at the Centre.
- Information on all aspects of transport and outdoor mobility for people with disabilities.

Walking made Easy!

Get the best out of life with Sungift's comprehensive range of electric vehicles, now including the New Sungift Legend Powerchair.

- No Licence required
- Indoor/Outdoor capability
- Highly manoeuvrable
- Superb kerb climbing ability
- Easily Transportable
- Available on Motability
- Unique Sungift 2 year warranty

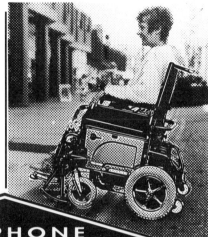

FREEPHONE
0800-626829

PHONE US NOW
OR FILL IN THE COUPON FOR
YOUR **FREE** COLOUR BROCHURE

NAME:

TEL:

ADDRESS:

POST CODE:

SMS
SUNGIFT MARKETING SERVICES

S.M.S. **FREEPOST** BF898
KEMPSTON BEDFORD MK42 7YF

SHOPMOBILITY SCHEMES

Shopmobility is a service which provides powered and manual wheelchairs and battery-operated scooters in shopping centres around the UK for all people with a mobility difficulty. It is usually free, although some schemes require a deposit. The service is open to anyone, young or old, whether their disability is temporary or permanent, and is available also for those with sport or accident injuries. You do not have to be registered disabled to use a scheme.

Shopmobility links together three stages of shopping: car access to designated parking, transfer from car to wheelchair, and wheelchair/scooter access to shops. The hours of operation normally match those of the shops. The schemes are often based in a car park with swift and safe access to shops, and normally comprise an office, reception area and wheelchair storage/work area. They are staffed by a full-time co-ordinator and a team of trained volunteers. Escorts can be arranged if needed, and many schemes have trained guides to assist people with visual or hearing impairments.

■ THE NATIONAL FEDERATION OF SHOPMOBILITY (NFS)

The NFS is a voluntary body and registered charity which aims to bring together and represent the interests and aspirations of the operators of Shopmobility schemes. It is an acknowledged first port of call for any organisation or individual considering establishing a scheme, offering free impartial advice and information about any aspect of starting up.

The NFS publishes a bi-annual directory which lists all Shopmobility schemes, their contact address and telephone number, together with their opening hours, and what sort of wheelchairs they have.

For more information on any aspect of Shopmobility you should contact:
Isobel Bracewell
New Enquiries Co-ordinator
National Federation of Shopmobility
80 Hilton Street, Aberdeen AB2 3QS
☎ (01224) 484957

Below are the Shopmobility schemes in operation at the time of writing – new schemes are opening all the time. Some of those listed were in an embryonic stage as we went to press, and their details were not complete. If you want to know more, contact the NFS. A key has been used to indicate what equipment is available:

    M = manual wheelchair
    P = powered wheelchair
    S = powered scooter

ABERDEEN Flourmill Lane Car Park, Aberdeen AB9 1EZ ☎ 01224-630009 Open Tues to Thurs 10am-4pm, Fri 8.30am-4pm, Sat 8.30am-1pm
   MPS

ARNOLD Croft Road Car Park, Arnold, Nottingham ☎ [to be arranged]
Open Mon, Wed and Fri 9.30am-12.30pm and 1pm-4pm
   MPS

BARKING AND DAGENHAM Centre Management, Vicarage Field Shopping Centre, Barking, Essex IG11 8DQ ☎ 0181-507 7773
Open Tues to Sat 10am-4pm
   MPS

BARROW IN FURNESS Shopmobility Furness, 11 The Mall Barrow in Furness, Cumbria LA14 1HL ☎ 01229-434039
Open Mon to Sat 10am-4pm
   MPS

BASILDON 103A Lower Galleries, Eastgate Centre, Basildon, Essex SS14 1AG
☎ 01268-533644 Open Mon to Fri 9am-4pm
   MPS

BATLEY
28 Alfreds Way, Batley, West Yorkshire WF17 5DR
☎ 01924-422417 Open Mon to Fri 9am-5pm
   MS

BATH 4 Railway Street, Bath, Avon BA1 1PG
☎ 01225-481744, Minicom: 01225-481773
Open Tues to Fri 9.30am-5.30pm, Sat 9am-1pm
   MPS

BEDFORD 1 The Howard Centre, Bedford, Bedfordshire, MK40 1QA ☎ 01234-348000
Open Mon to Sat 9.30am-4.30pm
   MPS

BEXLEYHEATH 1st Floor, Norwich Place End, Broadway Shopping Centre, Bexleyheath, Kent DA6 7JJ ☎ 0181-301 5237
Open Tues and Fri 10am-3pm
   MPS

BIRKENHEAD 5 St John Street, Birkenhead, Merseyside L41 6HY ☎ 0151-647 6162
Open Mon to Sat 9.30am-4.30pm
   MPS

BIRMINGHAM Birmingham Markets, Markets Customer Centre, Edgbaston Street, Birmingham
☎ 0121-643 4130 NB Correspondence to Birmingham City Council, Commercial Services Department, Manor House, 40 Moat Lane, Birmingham B5 5BD)
Open Mon to Fri 9.30am-4.30pm
   M

BRADFORD Unit 174, John Street Market, Rawson Road, Bradford, West Yorkshire BD1 3ST
☎ 01274-754076 Open Mon, Tues, Thurs and Fri 9am-5pm, Wed and Sat 9am-12 noon
   MPS

Shopmobility Schemes

Braintree   The Old Fire Station, Drury Lane, Swanside, Braintree, Essex CM7 7UW
☎ 01376-346535   Open Mon to Sat 9am-4.30pm   S

Brentwood from early 1995-   Cockfield Road Multi-storey Car Park, Brentwood, Essex CM14 4BN   ☎ 01277-219987
Open Mon to Sat 8am-9pm   MS

Brierley Hill   Merryhill Shopping Centre, Brierley Hill, West Midlands DY5 1SY
☎ 01384-481141   Open Mon, Tues, Wed and Fri 10am-8pm, Thurs 10am-9pm, Sat 9am-7pm   MPS

Bromley   Glades Shopmobility, The Glades Car Park, High Street, Bromley, Kent BR1 1DN
☎ 0181-313 9292   Open Mon to Wed, Fri and Sat 9am-5pm, Thurs 9am-8pm   MPS

Burton upon Trent   Unit 35, Octagon Shopping Centre, Park Street, Burton upon Trent, Staffordshire DE14 3TN
☎ 01283-515191   Open Mon to Fri 9.30am-4.30pm   MPS

Bury   Cross Street Car Park, Bury, Lancashire BL9 0PE   ☎ 01601-764996
Open Wed to Fri 9.30am-4.30pm, Sat 9.30am-3pm, Tues limited service 9.30am-3pm   MP

Cambridge   Level 9, Lion Yard Car Park, Cambridge   ☎ 01223-463370
NB Correspondence to Cambridge City Council, Engineers Department, The Guildhall, Cambridge)   Open Mon to Sat 10am-4pm   MPS

Cardiff   Oxford Arcade Multi-storey Car Park, Bridge Street, Cardiff, South Glamorgan CF2 2EB
☎ 01222-399355   Open Mon to Sat 9am-5pm   MPS

Carlisle   The Lanes Car Park Level 2), Carlisle, Cumbria   ☎ [to be arranged]
Open provisional-Mon to Sat 10am-5pm   MPS

Chelmsford   Cleansing Depot, Shelwater, Baddow Road, Chelmsfod CM2 7RB
☎ 01245-353771NB Permit holders only. Permits available from the Environmental Health Department, Civic Centre, Chelmsford, Essex CM1 1JEOpen Mon to Sat 9am-5.30pm   MPS

Cheltenham   Level P1, Beechwood Place Shopping Centre, Cheltenham, Gloucestershire GL50 1DQ,   ☎ 01242-255333
Open Mon to Sat 9.30am-4.30pm   MPS

Chesham   The Malthouse, Elgiva Lane, Chesham, Buckinghamshire HP5 2JD
☎ 01494-778400   Open Mon to Fri 9am-1pm   S

Chesterfield   Ground Floor, Multi-storey Car Park, New, Beetwell Street, Chesterfield, Derbyshire S40 1QR   ☎ 01246-209668
Open Mon to Sat 9.30am-4.30pm   MPS

Colchester   15 Queen Street, Colchester, Essex CO1 2PH   ☎ 01206-369099
Open Tues to Fri 10am-4pm   MPS

Coventry   The Barracks Car Park, Upper Precinct, Coventry, CV1 1DD
☎ 01203-832020   Open Mon to Sat 9am-5pm   MPS

Cowley   Templars Square Shopping Centre 129 Pound Way, Cowley, Oxford OX4 3XH
☎ 01865-748867   Open Tues and Fri 9am to 5pm   MPS

Clacton-on-Sea   Tendring Dial, 62 Station Road, Clacton-on-Sea, Essex CO15 1SP
☎ 01255-435566   Open Mon to Fri 10am-4pm   PS

Croydon   Whitgift Car Park, Wellesley Road, Croydon, Surrey   ☎ 0181-688 7336
NB Wheelchairs are available also from East Croydon BR station, the Drummond Centre and Surrey Street Car Park.
Open Mon to Sat 9am-5pm   MPS

Cwmbran   32 Gwent Square (opposite library), Town Centre, Cwmbran, Gwent NP44 1PL
☎ 01633-862951
Open Mon to Fri 9am-5pm, Sat 10am-3pm   MPS

Darlington   20/22 Horsemarket, Darlington, Co Durham DL1 5PT   ☎ 01325-461496
Open Mon to Sat 10am-5.30pm   MPS

Dartford   Priory Shopping Centre, Dartford, Kent, DA1 2HS   ☎ 01322-220915
Open Mon, Tues and Sat 8am-6pm, Wed, Thurs and Fri 8am-8pm   M

Derby   Bold Lane Car Park, Derby DE1 3NT
☎ 01332-200320   Open Mon-Fri 9.30am-5pm   MPS

Dewsbury   Level Best, Town Hall Way, Dewsbury, West Yorkshire WF12 8EQ
☎ 01924-455149
Open Mon to Fri 9am-5pm   MPS

Doncaster   Lower Mall Frenchgate Centre, St Sepulchre Gate, Doncaster, South Yorkshire,
☎ 01302-760742
NB Correspondence to Doncaster Community Transport, 46A Broxholme Lane, Doncaster DN1 2LN
Open Mon to Sat 9.30am-4.30pm   MS

DUMFRIES Holywood Trust Building, Old Assembly Close, Dumfries DG1 2PH
☎ 0345-090904 Open Tues to Fri 10am-4pm   **MPS**

DUNDEE Gellatly Car Park, Dundee
☎ 01382-228525 Open Tues to Thurs 10am-3pm   **MS**

EASTERHOUSE Unit 6, 7 Ware Road, Easterhouse, Glasgow, G34 9AB ☎ 0141-773 3356 Open Mon to Fri 9am-4.30pm   **MP**

EDINBURGH AND LOTHIAN The Mound Centre, Edinburgh EH2 ☎ 0131-225 9559
NB Correspondence to LRC, King Stables' Yard, King Stables' Road, Edinburgh EH2 2YJ
Open Tues to Sat 10am to 4pm
also at: Gyle Shopping Centre ☎ 0131-225 9553
Open Mon to Thurs 10am to 6pm   **MPS**

FALKIRK Level 4, Howgate Shopping Centre, High Street, Falkirk FK1 1DN ☎ 01324-611770
Open Mon, Tues and Thurs 10.30am-4pm
also at: Level 4, Callendar Square Car Park
☎ 01324-630500
Open Wed and Fri 10.30am-4pm   **MPS**

GATESHEAD Metro Centre, Gateshead, Tyne & Wear, NE11 9YG, ☎ 0191-460 5299
Open Mon to Wed and Fri 10am-8pm, Thurs 10am-9pm, Sat 9am-7pm   **M**

GLOUCESTER Hampden Way, Gloucester
☎ 01452-396898 NB Correspondence to The Herbert Warehouse, The Docks, Gloucester GL1 2EQ Open Mon to Sat 9.30am-5pm   **MPS**

HAMILTON 14 Lamb Street, Hamilton, Strathclyde ML3 6AH ☎ [to be arranged]
Open Mon to Fri 9.30am-4pm   **MPS**

HARLOW Post Office Road Car Park, The High, Harlow, Essex CM20 1BJ ☎ 01279-446188
Open Mon to Sat 9.15am-4pm   **MPS**

HARROW St Annes Shopping Centre, St Annes Road, Harrow, Middlesex HA1 1AR
☎ 0181-861 2282 Open Wed 10am-4pm   **MS**

HATFIELD 98 Town Centre, The Commons, Hatfield, Hertfordshire AL10 0NG,
☎ 01707-262731
also at: The Galleria Shopping Centre)
Open Tues to Sat 10am-5pm   **MPS**

HEMEL HEMPSTEAD Level A Car Park, Marlowes Shopping Centre, Marlowes, Hemel Hempstead, Hertfordshire HP1 1DX ☎ 01442-259259
Open Mon to Sat 9.30am-5pm   **MPS**

HEREFORD Maylord Orchards Car Park, Blueschool Street, Hereford HR4 9EU
☎ 01432-342166 NB Correspondence to Nikki Gough, City Housing and Works Department, Garrick House, Widemarsh Street, Hereford HR4 9EN Open Mon to Sat 9am-5pm   **MPS**

HIGH WYCOMBE Level 2, Newland Multi-storey Car Park, Newland Street, High Wycombe, Buckinghamshire HP11 2JD ☎ 01494-472277
Open Mon to Sat 10am-4pm   **MPS**

HUDDERSFIELD Level Best, Disability Access Point The Day Centre, Zetland Street, Huddersfield HD1 2RA ☎ 01484-453000
Open Mon to Fri 9am-5pm   **MPS**

IPSWICH Buttermarket Centre, St Stephen's Lane car access via Falcon Street), Ipswich, Suffolk, IP4 1HU ☎ 01473-222225
Open Mon to Sat 9.30am-5pm   **MPS**

KEIGHLEY Cooke Street, c/o Keighley Town Hall, Bow Street, Keighley, West Yorkshire BD21 3PA,
☎ 01535-618225
Open Mon and Wed to Fri 9am-5pm, Tues and Sat 9am-1pm   **MPS**

KETTERING (from January) 1995 The Old Ambulance Station, Market Street, Kettering, Northamptonshire ☎ [to be arranged]
Open Mon to Fri 9am-5.30pm, Sat 9am-12.30pm   **MPS**

KINGSTON UPON THAMES Eden Walk Car Park, Union Street, Kingston upon Thames, Surrey KT1 1BL ☎ 0181-547 1255
Open Mon to Sat 10am-5pm   **MPS**

KIRKCALDY
Mercat Centre Car Park, Tolbooth Street, Kirkcaldy KY1 1NJ ☎ 01592-640940,
Open Tues to Sat 10am-5pm   **MPS**

LEEDS 61 Vicar Lane, Leeds LS1 6BA
☎ 0113-246 0125
Open Mon to Sat 9.30am-4.30pm   **MPS**

LEICESTER Level 2, The Shires Car Park, High Street, Leicester ☎ 0116-253 2596
Open Mon to Fri 10am-5pm, Sat 10am-1pm   **MPS**

LEIGH Leigh Market Hall, Spinning Gate Shopping Centre, Leigh, Lancashire WN7 4PG,
☎ 01942-683163 Open Mon to Sat 10am-4pm   **MPS**

LEWISHAM Lewisham Centre, 3rd floor of car park, Lewisham, London SE13
☎ 0181-698 3775 Open Tues 10am-4pm   **MS**

LICHFIELD Multi-storey Car Park, Castle Dyke, Lichfield, Staffordshire WS1B 6RB
☎ 01543-254213 Open Mon-Fri 10am-4pm
MPS

LINCOLN Tentercroft Street Car Park, Lincoln
☎ 01522-544983 Open Mon-Sat 9am-4pm M

LIVERPOOL Clayton Square, Liverpool L1 1QR
☎ 0151-708 9993 Open Mon to Sat 9am-5pm
MPS

LUTON Level 3, Market Car Park, Melson Street, Luton, Bedfordshire LU1 2LJ ☎ 01582-38936, NB Correspondence to Arndale Management Office, Unit 37, Arndale Centre, Luton), Open Mon to Sat 9.30am-4.30pm MPS

MAIDSTONE King Street Multi-storey Car Park Maidstone, Kent ME14 1EN ☎ 01622-755759
Open Mon to Sat 7.45am-6pm MS

MANCHESTER Old Smithfield Market, Swan Street, Manchester M4 5JZ ☎ 0161-839 2552
Open Mon to Sat 9am-5pm PS

MEXBOROUGH Dearne Valley Community Transport, Bus Station, Johns Street, Mexborough S64 9HS ☎ 01709-571693
Open Mon to Fri 10am-2.30pm MPS

MILTON KEYNES Shopping Information Centre, Midsummer Arcade, Secklow Gate West, Central Milton Keynes, Buckinghamshire MK9 3ES
☎ 01908-678641
Open Mon to Wed 9.30am-5.30pm, Thurs and Fri 9.30am-7.30pm, Sat 9am-5.30pm MP

NELSON Town Hall Reception, Nelson, Lancashire ☎ 01282-602000
Open Wed and Fri 9am-4pm MPS

NEWBURY (FROM APRIL 1995)
Ground Floor, Northbrook Car Park, Pembroke Road, Newbury, Berkshire ☎ 01635-42400 ext 2510 (Shopmobility number – to be arranged)
Open Mon to Sat 10am-4pm MS

NEWCASTLE UPON TYNE
Centre Manager's Office, Eldon Square Shopping Centre, Eldon Court, Percy Street, Newcastle upon Tyne NE1 7JB
☎ 0191-261 1891/1714 Open Mon to Wed and Fri 9am-5.30pm, Thurs 9am-8pm, Sat 9am-6pm
M

NEWPORT 193 Upper Dock Street, Newport, Gwent NP9 1DA ☎ 01633-223845
Open Mon to Sat 9am-5pm MPS
also at: Kingsway Shopping Centre
☎ 01633-243688

NEWTON ABBOT Multi-storey Car Park, Sherbourne Road, Newton Abbot, Devon
☎ 01626-335775 Open Wed and Fri 10am-4pm
MPS

NORTHAMPTON Greyfriars Car Park, Greyfriars, Northampton ☎ 01604-233714, NB Correspondence to 13 Hazlewood Road, Northampton NN1 1LG
Open Tues to Fri 10am-4pm MPS

NORWICH 2 Castle Mall, Norwich, Norfolk NR1 3DD ☎ 01603-766430
Open Mon to Sat 10am-5pm MPS

NOTTINGHAM St Nicholas Centre, Stanford Street, Nottingham NG1 6AE ☎ 0115-958 4486
Open Mon to Sat 9am-5pm MPS

PETERBOROUGH Level 11, Queensgate Car Parking Complex, Peterborough, Cambridgeshire PE1 1NT ☎ 01733-313133
NB Correspondence to PCVS, 51 Broadway, Peterborough PE2 0EY)
Open Mon to Wed and Fri 10am-5pm, Thurs 10am-7.30pm, Sat 9am to 5pm MPS

PLYMOUTH Charles Cross Car Park, Plymouth, Devon, PL1 1AT ☎ 01752-600633
Open Mon to Sat 9am-4.30pm MPS

POOLE Level B, Multi-storey Car Park, Kingland Road Service Road, Poole, Dorset BH15 1TA
☎ 01202-661770 Open Mon-Sat 10am-4pm
MPS

PRESTON 28 Friargate, Preston, Lancashire PR1 3ALL ☎ 01772-204667
Open Mon to Fri 9am-5pm MPS

REDBRIDGE Ground Level, The Exchange Car Park, High Road, Ilford, Essex IG1 1RS
☎ 0181-478 6864
Open Mon to Fri 10am-4pm MPS

REDDITCH Car Park 3 (Access 3), Kingfisher Centre, Redditch, Hereford & Worcester B97 4HL ☎ 01527-69922
Open Mon to Wed and Fri 9am-6pm, Thurs 9am-7.30pm, Sat 9am-5.30pm MPS

RETFORD Chancery Lane Car Park, Retford, Nottinghamshire ☎ 01777-705432
Open Tues to Sat 9am-4.30pm MPS

ROCHDALE Unit 3, Bus Station Concourse, Smith Street, Rochdale, Greater Manchester OL16 1YG
☎ 01706-865986
Open Mon and Wed to Fri 10am-4pm, (Sat by arrangement) MPS

ROTHERHAM Scooter Loan, Community Transport Handybus, Erskine Road, Rotherham, South Yorkshire S65 1RF
☎ 01709-820792/820760
Open Mon to Fri 9am-5pm MS

SANDWELL The Ground Floor, Multi-storey Car Park, Sandwell Shopping Centre, West Bromwich B70 7NJ ☎ 0121-553 1943
Open Mon to Wed 9am-5pm, Thurs and Fri 9am-5.30pm, Sat 8.30am-4.30pm MPS

SCUNTHORPE John Street, Scunthorpe, Humberside DN15 6RB ☎ 01724-289636,
Open Tues to Sat 9.30am-5pm MPS

SHEFFIELD Meadowhall Shopmobility, Meadowhall Centre Ltd, Management Suite, 1 The Oasis, Meadowhall Centre, Sheffield S9 1EP
☎ 0114-256 8800
Open Mon-Thurs 9am-8pm, Fri 9am-9pm, Sat 9am-7pm, Sunday 11am-6pm MS

SOUTHEND ON SEA Farringdon Car Park, off Elmer Approach, Southend on Sea, Essex SS1 1NE ☎ 01702-339682
Open Tues and Thurs to Sat 10am-4pm MP

STEVENAGE 15 Queensway, Stevenage, Hertfordshire ☎ 01438-350300
Open Wed, Thurs, Fri and Sat 10am-4pm MPS

SUTTON 3rd Floor, St Nicholas Centre Car Park, St Nicholas Way, Sutton, Surrey SM1 1AY
☎ 0181-770 0691
Open Mon to Fri 10am-4pm MPS

SUTTON COLDFIELD Gracechurch Shopping Centre, Sutton Coldfield, West Midlands B72 1PH ☎ 0121-355 1112
Open Mon to Sat 9am-5pm M

TELFORD Red Oak Car Park, Telford Shopping Centre, Telford, Shropshire TF3 4BX
☎ 01952-291370 ext 214
Open Mon to Thurs and Sat 9am-5pm, Fri 9am-8pm MPS

THURROCK Thurrock Lakeside Shopping Mall, West Thurrock Way, West Thurrock, Grays, Essex RM16 1WT ☎ 01708-869933 ext 566
Open Mon to Fri 10am-7pm MS

WARRINGTON Birchwood Shopping Mall, Wellington, Warrington, Cheshire WA3 7PG
☎ 01925-822411
Open Mon to Sat 10am-4pm MPS

WATFORD Level 4, Charter Place Car Park, Watford, Hertfordshire WD1 2RN ☎ 01923-211020 Open Tues to Sat 10am-5pm MPS

WIGAN 1 Wigan Gallery, The Galleries Shopping Centre, Wigan, Greater Manchester WN1 1AR
☎ 01942-825520
Open Mon to Sat 10am-4pm MPS

WOKING Level One, The Peacocks, Victoria Way, Woking, Surrey GU21 1GD ☎ 01483-776612,
Open Mon to Fri 9.30am-5pm,Sat 9am-5pm MPS

WOLVERHAMPTON 12 Cleveland Street, Wolverhampton, West Midlands WV1 3HH
☎ 01902-773190,
Open Mon to Fri 9.30am-5pm, Sat 9.30am-4.30pm MPS

WORCESTER 54 Friary Walk, Crowngate Centre, Worcester WR1 3LE ☎ 01905-610523,
Open Mon to Sat 9am-5pm, bank holidays 10am-4pm MPS

WORKSOP Priory Centre, off Bridge Place, Worksop, Nottinghamshire S80 1DG
☎ 01909-479070
Open Mon 9am to 3pm, Wed-Sat 9am-4.30pm MPS

WORTHING c/o United Reformed Church, Shelley Road, Worthing, West Sussex BN11 1TT
☎ 01903-820980
Open Tues to Sat 9am-5pm MPS

YEOVIL Petter House, Petter Way, Yeovil, Somerset , BA20 1SH ☎ 01935-75914
Open Mon to Fri 9am-4.30pm M

WATFORD Level 4, Charter Place, Watford, Hertfordshire WD1 2RN ☎ 01923-211020
Open Tues to Sat 10am-5pm MPS

WIGAN 1 Wigan Gallery, The Galleries Shopping Centre, Wigan, Greater Manchester WN1 1AR
☎ 01942-825520

Facilities at Supermarkets

The big supermarket chains listed below were asked to describe the general facilities provided for disabled people at their stores. The signs are that they are all trying to create better access and make shopping easier, particularly in any new stores, especially the latest out-of-town complexes. Older high street branches may not be as easy to get around or use. If you think your local store could do more to help, it is worth having a word with the store manager or contacting the relevant customer services department.

■ ASDA

Car parking spaces are available adjacent to store entrances. The stores are all on one level, lifts are provided to the shop floor.

Trolleys are available which can be attached to wheelchairs. Stores have low-level shelving to make movement and selection of goods easier. Extra-wide checkout aisles are provided and are clearly marked. Staff will be able to help at all times, ask at the customer service desk.

People in wheelchairs can be accommodated in customer restaurants. Telephones are positioned at a low level on the wall.

For more information call 0113-243 5345 and ask for the Customer Services Department.

■ BUDGENS

All the newer stores have disabled toilets, wide aisles, automatic doors and a wide checkout. Car parks have a section with disabled parking places close to the store entrance.

For more information, contact the Personnel Department at Budgens on 0181-422 9511.

■ SAFEWAY

The stores which have car parks provide parking spaces for disabled people close to the shop entrance. Special trolleys are provided to attach to wheelchairs. Checkouts and counters are built to a height that is accessible from a wheelchair and the most popular goods are placed on the most easily reached shelves. There is a free bag-packing and carry-out service. The vast majority of stores have disabled toilets. A free taxi service is also available.

For more information call Safeway Customer Relations on 01622-712000.

■ SAINSBURY'S

All Sainsbury's store car parks have spaces reserved close to the store entrance for disabled badge holders. All modern stores have automatic entry and exit doors. Some Sainsbury's stores have motorised wheelchair trolleys ('Care Carts')

for disabled customers (ask at your local store).

All stores have adapted trolleys for use with wheelchairs. Some have trolleys with special padded seats for disabled children.

Customer Service Assistants are available on request to help customers at checkouts by packing goods and taking shopping to the car park. Most stores have two wide checkouts for wheelchair users, clearly indicated by the wheelchair symbol.

Where stores have toilets, they are accessible for disabled people.

If you want more information, call the Customer Services helpline free on 0800 636262.

■ SUMMERFIELDS

Where there are car parks, spaces are provided for disabled drivers close to the store entrance. If customer toilets are available, one is accessible for disabled people. The stores have one wide checkout, and a bag-packing and carry-out service operates. Special trolleys can be attached to wheelchairs.

For more information call the head office on 0117 935 9359.

■ TESCO

Where a store has a car park, spaces are provided for disabled people close to the store entrance. There are dropped kerbs for wheelchairs, and road crossings with lights make it easier for disabled people to approach the store.

If a store has a revolving door, there is a door at the side which can be opened automatically with a push button to allow wheelchair access. Electronic gates in stores are accessible by wheelchair. Special trolleys are provided for disabled customers, and in the produce areas bags are placed low enough for those in wheelchairs to reach. Staff are available to help with shopping if a customer is blind, has poor eyesight or is in a wheelchair. Wide checkouts are provided.

If a store has toilets, access will be possible for disabled people.

For more information call Tesco Customer Relations free on 0800 505555.

■ WAITROSE

Where there are customer car parks, spaces are available for disabled customers close to the shop entrance. The stores have wide checkouts and special trolleys that can be clipped to a wheelchair. If a store has customer toilets, one is provided for disabled people.

For more information call the Customer Services Department on 01344 424680.

The Royal Association for Disability and Rehabilitation (RADAR)

12 City Forum, 250 City Road, London EC1V 8AF ☎ 0171-250 3222, Minicom: 0171-250 4119

RADAR is a national organisation working with and for physically disabled people. The Association's policy is to remove architectural, economic and attitudinal barriers which impose restrictions on disabled people, and to ensure that disabled people are able to play their full role in the community. RADAR is particularly involved in the areas of education, employment, mobility, social services, housing and social security. The Association operates in conjunction with an affiliated network of around 500 local and national organisations.

RADAR also produces publications on a wide range of subjects, including a number on mobility, travel and holidays. Below are listed some which may be of interest to RAC Members.

■ MOBILITY AND TRAVEL

Seatbelts and Disabled People
A survey conducted by RADAR for the Department of Transport on the design and suitability of seatbelts for use by people with disabilities.

Parking Concessions for Disabled People
Report on a survey of local authority parking concessions for people with disabilities. It includes examples of good practice and RADAR's parking recommendations.

Parking Concessions Offered to Orange Badge Holders at Off-street Car Parks Operated by District Councils in England

The following are Mobility Factsheets, giving concise information on a wide range of topics:
- Motoring with a wheelchair
- Exemption from Vehicle Excise Duty (VED)
- Relief from VAT on motor vehicles
- Driving licences
- Assessment centres and driving instruction
- Insurance
- Motoring accessories
- Cash help for mobility needs
- Discounts and concessions available to disabled people on the purchase of cars and other related items
- Car control manufacturers, suppliers and fitters.

■ HOLIDAYS

Holidays in the British Isles – A Guide for Disabled People
This annual guide to accommodation and facilities in the UK includes chapters on planning and booking a holiday, voluntary organisations and commercial companies involved in holiday provision for disabled people, activity holidays, transport and useful publications, along with full regional listings of suitable holiday accommodation. The guide also provides useful regional addresses and ideas for places to visit.

Holidays and Travel Abroad –
A Guide for Disabled People
This annual guide to accommodation and facilities abroad gives basic holiday details of around 100 countries and advice on the best organisations to contact for more details. There are chapters on insurance, transport, voluntary organisations and commercial companies involved in holiday provision for disabled people, international hotel companies and reservation agencies.

The following are Holiday Factsheets:
- Useful addresses and publications for the disabled holidaymaker
- Sport and outdoor activity holidays
- Holiday insurance cover
- Holiday finance
- Planning and booking a holiday
- Holiday transport
- Escort, taxi and private ambulance services
- Equipment for hire.

■ SPORTS AND LEISURE

Country Parks
This access guide gives information on facilities such as parking, toilets and catering for disabled people wishing to use parks throughout the UK.

Spectator Sports – A Guide for Disabled People
The guide looks at sports such as football, athletics, basketball, cricket, ice hockey, motor sport, rugby, tennis, and horse and greyhound racing. It covers the facilities and access for disabled people at many sporting venues and gives the information needed to plan visits to watch events.

Historic Buildings of England –
A Guide for Disabled People
Hundreds of historic houses, churches, castles and cathedrals are covered in this guide. Parking arrangements and internal and external access details are given, along with the availability of facilities such as wheelchairs for loan, accessible toilets and staff assistance offered. A description is also given of each property, together with the property's other attractions, such as gardens, gift shops and restaurants.

For a complete publications list and an order form contact RADAR at the above address.

NATIONAL KEY SCHEME (NKS)

A standard lock is fitted to accessible public conveniences which have had to be locked to prevent vandalism and mis-use. The scheme offers independent access for disabled people to around 4,000 locked public toilets around the UK. You can obtain keys from the local authority or direct from RADAR, at a cost of £2.50, if the order includes the name and address of the disabled person with a declaration of disability. Without the declaration RADAR has to charge VAT on the £2.50. At the time of writing, a book listing all the NKS toilets was in preparation and should be available from RADAR in March 1995.

DIAL UK

Park Lodge, St Catherine's Hospital, Tickhill Road, Balby, Doncaster DN4 8QN
☎ 01302 310123

DIAL UK is the national association for the DIAL network of over 100 disability information and advice services. DIAL groups give free, independent and impartial advice and are run and staffed by people with direct experience of disability. DIAL UK itself provides an information service supplying reference data on all aspects of disability to DIAL groups and other organisations.

ROUTE FINDING AND ACCESS MAPS PROJECT (RAMP)

c/o Disabled Motorists' Federation, National Mobility Centre, Unit 2A, Atcham Estate, Shrewsbury SY4 4UG ☎ 01743 761181

RAMP is a free service for disabled drivers and drivers with disabled passengers. The aim is to help disabled drivers to find petrol filling stations that are not self-service, stepless cafés and restaurants with parking at the door, toilets built especially for disabled users, overnight accommodation with no steps or stairs, and wheelchair-friendly places to visit.

The service offers outline maps of the best routes between places chosen by users, detailed maps of urban areas, lists of directions, lists of addresses, and information about places marked by symbols on the maps. Maps are in colour and show attended-pump filling stations, RADAR key-scheme toilets, and wheelchair-accessible cafés, B&Bs and tourist attractions.

All you have to do is call the MAPLINE number – 01743-761181 – and state the starting point, destination and setting-out or arrival time. Maps will then be posted, usually within 24 hours.

HEMCO

Electrically Operated Bed System

Tailoring Chairs and Beds towards people's needs.

HEMCO Products conform to BS 5852 WOODCRIB 5 (Fire Retardancy)
BS 3456 (Electrical Safety) BS 4875 (Stength and Stability).
HEMCO Products carry a two year guarantee

HEMCO
UK Limited

Unit 59, Llandon Industrial Estate, Cowbridge, South Glamorgan Tel.: 01446 773394 Fax: 01446 772226

Travel at Home and Abroad

MOTORWAY SERVICE AREAS

Below is a list of all motorway service areas, with brief details of the siting of each. The vast majority in the UK are controlled by a small group of companies: Granada, Welcome Break, Pavilion, RoadChef and Blue Boar. All service areas provide specific parking spaces for disabled drivers, and almost all have disabled access to toilets, refreshment areas, telephones and shops. Most service areas provide some kind of help for disabled people at the petrol station. Where we know that the Servicecall facility is provided at a service area, this is mentioned. If we are aware of any special facilities or access problems, these are also mentioned.

The accommodation available is stated at the end of an entry (see also the next section on lodge accommodation).

A1(M)

DURHAM – at A177 Bowburn junction (RoadChef) Tel: 0191-377 9222. Servicecall.
RoadChef Lodge Tel: 0191-377 3666.

WASHINGTON – Birtley bypass (Granada) Tel: 0191-410 3436. N and S bound sites, footbridge link. Servicecall.
Granada Lodge (S bound) Tel: 0191-410 0076.

A1(M)/A614

BLYTH (Granada) Tel: 01909-591841. Single site, off roundabout. Servicecall.
Granada Lodge Tel: 01909-591841.

M1

J2 SCRATCHWOOD (Welcome Break) Tel: 0181-906 0611. N and S bound single site by vehicle bridge. Cash dispenser (Barclays/Nat. West). Poor access (steep ramp, heavy doors). Key access to toilets, no access to phone.
Welcome Lodge Tel: 0181-906 0611.

J11–12 TODDINGTON (Granada) Tel: 01525-878400. N and S bound sites, footbridge link. S bound site has first-floor restaurant, accessible by service lift with staff help. Servicecall.
Granada Lodge (S bound) Tel: 01525-875150.

J14–15 NEWPORT PAGNELL (Welcome Break) Tel: 01908-217722. N and S bound sites, footbridge link. Cash dispenser (Barclays/Nat. West). N bound access poor (ramp, heavy doors). No access to phone.
Travelodge (N bound) Tel: 01908-610878.

J15–16 ROTHERSTHORPE (Blue Boar) ☎ 01327-78811. N and S bound sites, footbridge link. Lift at S bound site. Servicecall.

J16–17 WATFORD GAP (Blue Boar) ☎ 01327-78811. N and S bound sites, footbridge link. Servicecall.

J21–22 LEICESTER FOREST EAST (Welcome Break) ☎ 0116-238 6801. N and S bound sites. Cash dispenser (Link/Nat. West).

J22 off A50 LEICESTER MARKFIELD (Granada) ☎ 01530-244777. Single site, access from M1 and A50. Servicecall.
Granada Lodge ☎ 01530-244237.

J25–26 TROWELL (Granada) ☎ 0115-932 0291. N and S bound sites, footbridge link. Servicecall.

J30–31 WOODALL (Welcome Break) ☎ 0114-248 6434. N and S bound sites, footbridge link. Cash dispenser (Nat. West).

J33 off A630/A6102 SHEFFIELD (Granada) ☎ 0114-239 9990. Single site, access from M1. Servicecall.
Granada Hotel ☎ 0114 253 0935.

J38–39 WOOLLEY EDGE (Granada) ☎ 01924-830371. N and S bound sites, no access across motorway. Servicecall.
Granada Lodge (N bound) ☎ 01924-830569.

MOTORWAY SERVICE AREAS

M2
J4–5 MEDWAY (Pavilion)
☎ 01634-233343. E and W bound sites, footbridge link. Servicecall.
Medway Pavilion (E bound) ☎ 01634-377337.

M3
J4A–5 FLEET (Welcome Break)
☎ 01252-621656. W and E bound sites, footbridge link. Cash dispenser (Barclays/Link). No access to shop at W bound site. Servicecall.
Travelodge (W bound) ☎ 01252-815587.

M4
J2–3 HESTON (Granada)
☎ 0181-574 7271. W and E bound sites, no access across motorway. Servicecall.
Granada Lodge (W bound) ☎ 0181-574 5875.

J13 OFF A34 NEWBURY/CHIEVELEY (Granada)
☎ 01635-248024. Single site on roundabout, access from both sides of M4 and A34. Servicecall.

J14–15 MEMBURY (Welcome Break)
☎ 01488-71881. W and E bound sites, footbridge link. Cash dispenser (Nat. West – W bound). Access poor (heavy doors, restaurant via lift) E bound. Access fair (heavy doors, restaurant via steep ramp) W bound.

J17–18 LEIGH DELAMERE (Granada)
☎ 01666-837691. W and E bound sites, footbridge link. Cash dispenser (Barclays/Link/Nat. West). Servicecall.
Granada Lodge (E bound) ☎ 01666-837691.

J21 OFF A403 SEVERN VIEW (Pavilion)
☎ 01454-632851. W and E bound, single-side site. Cash dispenser (Nat. West).
Severn View Pavilion ☎ 01454-633313.

J23 MAGOR (Granada)
☎ 01633-880111. Single-side site. Servicecall.
Granada Lodge ☎ 01633-880111.

J33/A4232 CARDIFF WEST (Pavilion)
☎ 01222-891141. Single site, access from M4 both sides and A4232. Cash dispenser (Lloyds).
Cardiff West Pavilion ☎ 01222-892255.

J36 OFF A4061 SARN PARK (Welcome Break)
☎ 01656-655332. Single site, direct from E bound side, W bound off J36. Access via spiral ramp. Key access to toilets. No access to phone.
Travelodge ☎ 01656-659218.

J47/A483 SWANSEA (Pavilion)
☎ 01792-896222. Single-side site at roundabout. (Cash dispenser Lloyds).
Swansea Pavilion ☎ 01792-894894.

J49 OFF A48/A483 PONT ABRAHAM (RoadChef)
☎ 01792-884663. W and E bound, single site. Cash dispenser (Lloyds/Nat. West/Bank of Scotland). Servicecall.

M5
J3–4 FRANKLEY (Granada)
☎ 0121-550 3131. N and S bound sites, no access across motorway. S bound site has first-floor restaurant, accessible by service lift with staff help. Servicecall.
Granada Lodge (S bound) ☎ 0121-550 3261.

J7–8 STRENSHAM (Kenning Motor Group)
☎ 01684-293004. N and S bound sites, footbridge link.

J13–14 MICHAELWOOD (Welcome Break)
☎ 01454-260631. N and S bound sites, footbridge link. Cash dispenser (Barclays/Nat. West). *Travelodge* (open winter 1994/95) ☎ 0800 850 950.

J19 OFF A369 GORDANO (Welcome Break)
☎ 01275-373624. N and S bound, single site, via roundabout. Access fair (ramp, heavy doors to refreshments). Access to toilet difficult. Access to shop restricted.
Travelodge ☎ 01275-373709.

J21–22 SEDGEMOOR N BOUND (Welcome Break)
☎ 01934-750730. Access to shop and toilets limited.
Travelodge (N bound) ☎ 01934-750831.
Sedgemoor S BOUND RoadChef
☎ 01934-750888. Servicecall.

J25–26 TAUNTON DEANE (RoadChef)
☎ 01823-271111. N and S bound sites, footbridge link. Cash dispenser (Barclays – S bound). Servicecall.
RoadChef Lodge (S bound) ☎ 01823-332228.

J30 AT A376 EXETER (Granada)
☎ 01392-436266. W and E bound. Servicecall.
Granada Hotel ☎ 01392-74044.

M6
J3–4 CORLEY (Welcome Break)
☎ 01676-40111. N and S bound sites, foot-bridge link. (Barclays/Link/Nat. West). Cash dispenser. Key access to toilets. No access to phone and restricted access to shop S bound.

35

J10A–11 Hilton Park (Pavilion)
☎ 01922-412237. N and S bound sites, footbridge link. Cash dispenser (Lloyds/Nat. West). RADAR key access to toilets. No help at petrol station.
Hilton Park Pavilion (S bound) ☎ 01922-414100.

J15–16 Keele (Welcome Break)
☎ 01782-626221. N and S bound sites, footbridge link. Cash dispenser (Barclays/Nat. West). Key access to toilets. Access to shop restricted S bound.

J16–17 Sandbach (RoadChef)
☎ 01270-767134. N and S bound sites, footbridge link. Cash dispenser (Barclays – S bound). Servicecall.

J18–19 Knutsford (Pavilion)
☎ 01565-634167. N and S bound sites, footbridge link. No access (steps) to main restaurant (access to fast food). Servicecall.

J27–28 Charnock Richard (Welcome Break)
☎ 01257-791494. N and S bound sites, footbridge link. Cash dispenser (Barclays/Link/Nat. West).
Travelodge (N bound) ☎ 01257-791746.

J32–33 Forton (Pavilion)
☎ 01524-791775. N and S bound sites, footbridge link. RADAR key access to toilets. Refreshments via two lifts N bound. Access to café S bound via road for N bound facilities. Servicecall.
Forton Pavilion (N bound) ☎ 01524-792227.

J35–36 Burton West) (Granada)
☎ 01524-781234. N bound only. Servicecall.

J36–37 Killington Lake (RoadChef)
☎ 01539-620739. S bound only. Servicecall.
RoadChef Lodge (S bound) ☎ 01539-621666.

J38–39 Tebay (Westmorland)
☎ 01539-624511. N and S bound.
Tebay Mountain Lodge (N bound) but access from S bound ☎ 01539-624351.

J41–42 Southwaite (Granada)
☎ 01697-473476. N and S bound sites, footbridge link. Servicecall.
Granada Lodge S bound) ☎ 01697-473131.

M11
J8 Birchanger service area under construction.

M23
J9/A23 Pease Pottage (Welcome Break)
☎ 01293-562852. Off motorway, at roundabout.

M25
J5–6 Clackett Lane (RoadChef)
☎ 01959-565577. E and W bound sites, no access across motorway. Servicecall.

J23 off A1(M) South Mimms (Welcome Break)
☎ 01707-646333. W and E bound, single site. Cash dispenser (Barclays/Link/Nat. West). Servicecall. *Travelodge* ☎ 01707-665440.

J30–31 Thurrock (Granada)
☎ 01708-865487. Single site, access N and S bound. Servicecall.
Granada Lodge ☎ 01708-891111.

M27
J3–4 Rownhams (RoadChef)
☎ 01703-734480. W and E bound sites, subway link. Servicecall.
RoadChef Lodge W bound) ☎ 01703-741144.

M40
J10 Cherwell Valley (Granada)
☎ 01869-346060. Single site, access N and S bound. Servicecall.
Granada Lodge (N bound) ☎ 01869-346111.

J12–13 Warwick (Welcome Break)
☎ 01926-651681. Single site, access N and S bound. Access to shops limited.
Travelodge (N bound) ☎ 01926-651681.

M42
J10/A5 Tamworth (Granada)
☎ 01827-260120. Single site, access from M42 and A5. Servicecall.
Granada Lodge ☎ 01827-260123.

M61
J6–8 Rivington Vale (Pavilion)
☎ 01204-68641. N and S bound, single site, vehicle bridge. Cash dispenser (Lloyds). Servicecall.

M62
J7–9 Burtonwood (Welcome Break)
☎ 01925-51656. W and E bound sites, subway link. Toilets in garage area. Access to shops limited.
Travelodge (W bound) ☎ 01925-710376.

J18–19 Birch (Granada)
☎ 0161-643 0911. W and E bound sites, footbridge link. Servicecall.
Granada Lodge (E bound) ☎ 0161-655 3403.

MOTORWAY AND A-ROAD ACCOMMODATION

J25–26 HARTSHEAD MOOR (Welcome Break)
☎ 01274-876584. W and E bound sites, footbridge link. Access E bound poor (steep ramp, heavy doors). Access W bound fair (steep ramp). Steep ramps to restaurants.
Travelodge (E bound) ☎ 01274-851706.

J33/A1 FERRYBRIDGE (Granada)
☎ 01977-672767. Single site – access from both sides of M62 and A1. Servicecall.
Granada Lodge ☎ 01977-672767.

■ SCOTLAND

M8

J4–5 HARTHILL (RoadChef)
☎ 01501-51791. W and E bound sites, footbridge link. Servicecall.

M74

J4–5 HAMILTON (RoadChef)
☎ 01698-282176. N bound only, single site. Servicecall.
RoadChef Lodge ☎ 01698-891904.

J5–6 BOTHWELL (RoadChef)
☎ 01698-854123. S bound only, single site. Servicecall.

A74(M)

GRETNA GREEN (Welcome Break)
☎ 01461-337567. N bound, access from both sides.
Travelodge (N bound) ☎ 01461-337566.

M80/M9

J9 STIRLING (Granada)
☎ 01786-813614. Single site, access from both sides of M80 and M9. Servicecall.
Granada Lodge ☎ 01786-815033.

M90

J6 KINROSS (Granada)
☎ 01577-863123. N and S bound sites. Servicecall.
Granada Lodge (N bound) ☎ 01577-864646.

■ A-ROAD SERVICES

Service areas on UK A-roads are dominated by Little Chef and Happy Eater, both owned by Forte. The standard opening hours for these two chains are 7am to 10pm all week. Disabled facilities are available at all restaurants, including a braille menu, ramps, adapted toilets and allocated parking spaces. For more details, and to get a copy of the Little Chef and Happy Eater brochures that list all the restaurants, phone 01256-812828.

MOTORWAY AND A-ROAD ACCOMMODATION

■ FORTE TRAVELODGES

These are located on motorways and on A-roads, linked either to a Welcome Break service area or a Happy Eater, Little Chef or Harvester restaurant. Each Travelodge has disabled parking spaces and at least one room with disabled facilities, including:
- level or ramped access
- immediate access to the room, on the ground floor, next to reception
- an alarm in the bedside unit, activating both a bell and light in the reception and adjacent restaurant
- a bedroom door fitted with an additional low-level spy-hole for use from a wheelchair
- remote control for TV/radio/alarm
- height of bed increased by three inches over the standard height for ease of access by wheelchair

All rooms have a double bed, a single sofa bed and a child bed, if needed, an en suite bathroom and shower, tea and coffee making facilities, feather duvets, colour TV and radio.

You can reserve a Travelodge room up to three months in advance, or request a brochure, by calling free on 0800 850 950. The line is open seven days a week 7am to 10pm. Booking by credit card will guarantee a room, no matter how late you arrive. Non-credit card bookings will be held until 6pm.

There are also Travelodges in France and the Republic of Ireland. You can get a brochure by calling the number above.

■ GRANADA LODGES

Most Granada Lodges are located on motorways, linked to service areas, with a few to be found on A-roads. Each bedroom has remote-control TV with Sky channels, direct-dial telephone, tea and coffee making facilities and en suite bathroom. Newspapers and hot and cold food can be delivered to the room. There is also the choice of smoking or non-smoking rooms. Adapted bedrooms for disabled people are available throughout the Granada Lodge network.

You can check in any time after 3pm up to 6pm, or any time at all if you book by credit card. To reserve a room, call free on 0800 555 300, 24 hours a day, seven days a week. A brochure is available detailing all the locations of Granada Lodges.

■ PAVILION LODGES

These are situated on motorways and on A-roads, connected to Pavilion service areas. Each lodge room has an en suite bathroom, colour TV with Sky channels, tea and coffee making facilities, alarm call service, hairdryer and trouser press. The reception areas are staffed 24 hours a day, so you can check in when you like. All Pavilion Lodges have two rooms adapted for disabled people, most with a low bath and rails in the bathroom, and an emergency button or cord in bathroom and bedroom.

To reserve a room or for more information call the reservation line free on 0800 515 836.

■ ROADCHEF LODGES

At the time of writing, all RoadChef Lodges are to be found on motorways, connected to RoadChef service areas. All rooms have private bathroom, colour TV (including satellite channels), radio, self-dial telephone, tea and coffee making facilities, hairdryer and trouser press. Each lodge has adapted rooms for disabled travellers.

RoadChef Lodges are open throughout the year, and you can make a booking free on 0800 834 719. To get more information, including a brochure detailing the locations of the lodges, call the above number or 01452-303373.

TOLL CONCESSIONS

Disabled people who use the following toll bridges and tunnels are entitled to concessions, as long as they meet certain requirements requested by the owners or operators.

In some cases, to qualify for a concession an application has to be made in advance to a bridge, tunnel or local authority – always check with the owners or operators first.

CLEDDAU BRIDGE (A477, DYFED)
☎ 01646-683517

Local drivers who are in receipt of mobility allowance or exempt from Vehicle Excise Duty cross free. If you qualify, you receive a book of 25 vouchers from the bridge office.

CLIFTON SUSPENSION BRIDGE (B3214, WEST OF BRISTOL)

Disabled drivers in receipt of mobility allowance or the higher-rate mobility component of the DLA can get an annual ticket for 50p from: The Bridge Master, Clifton Suspension Bridge, Leigh Woods, Bristol BS8 3PA

DARTFORD TUNNEL/DARTFORD BRIDGE (M25, KENT/ESSEX) ☎ 01322-221603

Northbound traffic uses the Dartford Tunnel, southbound traffic uses the Queen Elizabeth II Bridge. The crossing is free for those disabled drivers who are exempt from Vehicle Excise Duty.

DUNHAM BRIDGE (A57, LINCOLN TO WORKSOP)

Concessions are available only to those people driving vehicles supplied by a government department.

ERSKINE BRIDGE (M898/A82, STRATHCLYDE, NEAR GLASGOW) ☎ 0141-812 7022

Orange Badge holders cross free.

FORTH ROAD BRIDGE (A90/M90, LOTHIAN/FIFE, NEAR EDINBURGH) ☎ 0131-319 1699

Orange Badge holders cross free.

HUMBER BRIDGE (A15/A63, HUMBERSIDE, NEAR HULL) ☎ 01482-647164

Disabled drivers in receipt of mobility allowance or the higher-rate mobility component of the DLA cross free. An application for tickets must be made in advance to: Humber Bridge Board, Administration Building, Ferriby Road, Hessle, North Humberside HU13 0JG

ITCHEN BRIDGE (A3025, WOOLSTON TO SOUTHAMPTON) ☎ 01703-433449

Concessions are available for those using invalid carriages or for those disabled drivers who are exempt from Vehicle Excise Duty.

MERSEY TUNNEL

(A5139 and A5030, Liverpool to Birkenhead)
☎ 0151-236 8602

Concessions are granted to the following disabled people who are resident in the county of Merseyside:

- Orange Badge holders who drive and own their vehicle or who lease it through Motability
- disabled people who drive a vehicle supplied by a government department or who use their own vehicle for which they receive a government grant
- disabled drivers in receipt of War Pensioners' Mobility Supplement
- people who have been issued with an Orange Badge but are dependent on the vehicle owner, and who need to be driven for genuine reasons of disability. Both the Orange Badge holder and vehicle owner must live at the same address.

Further details can be obtained from: The General Manager, The Mersey Tunnels, George's Dock Way, Pier Head, Liverpool L3 1DD

SEVERN ROAD BRIDGE (M4, AVON/GWENT, NEAR BRISTOL/CHEPSTOW)
☎ 01454-632457/201102
Orange Badge holders cross free.

TAMAR BRIDGE (A38 PLYMOUTH TO LISKEARD)
Torpoint Ferry (A374 Plymouth to Torpoint)
☎ 01752-361577
Free vouchers are available for:
- recipients of mobility allowance
- those who receive the higher-rate mobility component of the DLA
- those who are exempt from Vehicle Excise Duty
- the registered blind.

An application for vouchers must be made in advance. Up to 100 vouchers per year may be issued after proof of entitlement and a passport-size photograph are supplied to the Joint Committee.
All enquiries must be accompanied by an s.a.e.

for an application form and the return of details. Full information can be obtained from:
The Tamar Bridge and Torpoint Ferry, Joint Committee, 2 Ferry Street, Torpoint, Cornwall PL11 2AX

TAY ROAD BRIDGE (A92 DUNDEE TO NEWPORT ON TAY) ☎ 01382-221881
Orange Badge holders cross free.

TYNE TUNNEL (A19) ☎ 0191-262 4451
Disabled drivers who are exempt from Vehicle Excise Duty cross free.

WHITCHURCH BRIDGE (B471 PANGBOURNE TO WHITCHURCH)
Orange Badge holders cross for free.

WHITNEY-ON-WYE BRIDGE (B4350 WHITNEY-ON-WYE TO HAY-ON-WYE)
Orange Badge holders cross for half-price.

PROVISION FOR DISABLED TRAVELLERS ON CAR FERRIES

The following ferry companies provide facilities for disabled travellers on their ferries and at ferry terminals. Some offer more facilities than others, often depending on the age of the vessels and on the length of the crossing, but it is always essential to make clear to a company what your particular requirements are, so that suitable arrangements can be made.

Please note that where a concessionary rate is available to members of either the DDA or the DDMC we advise you to contact those organisations direct before you book your crossing, to confirm the terms of the concession.

■ BRITTANY FERRIES
Millbay Docks, Plymouth PL1 3EW
☎ 01752-221321
The Brittany Centre, Wharf Road, Portsmouth PO2 8RU ☎ 01705-827701

UK TO FRANCE

PORTSMOUTH TO CAEN *Normandie*
A year-round service, with a crossing time of six hours and daily sailings – up to three returns daily in peak season.

PORTSMOUTH TO ST MALO *Bretagne*
A year-round service from mid-March, with a crossing time of eight and three-quarter hours and a daily return sailing from mid-March to early November – twice a week thereafter.

PLYMOUTH TO ROSCOFF *Quiberon*
A year-round service, with a crossing time of six

hours and up to three return sailings a day in peak season.

POOLE TO CHERBOURG *Barfleur*
A year-round service, with a crossing time of four and a quarter hours and two daily return sailings.

POOLE TO ST MALO *Duchesse Anne*
The service operates from mid-May to early October, with a crossing time of eight hours and four return sailings a week.

UK TO SPAIN

PLYMOUTH TO SANTANDER *Val de Loire*
The service operates from mid-March to November, with a crossing time of 24 hours and up to two return sailings a week.

PORTSMOUTH TO SANTANDER *Val de Loire*
The service operates from November to mid-March, with a crossing time of 29 to 30 hours and weekly return sailings.

IRELAND TO FRANCE

CORK TO ROSCOFF *Val de Loire, Duchesse Anne* and *Quiberon*
The service operates from March to October, with a crossing time of thirteen and a half hours and up to two return sailings a week.

CORK TO ST MALO *Val de Loire* and *Duchesse Anne*
The service operates from mid-April to mid-

September, with a crossing time of eighteen and a half hours and one return sailing a week.

There are cabins for disabled passengers on each of the ships in the Brittany Ferries fleet. *Quiberon* has the most, with a total of 11 cabins. These are adapted to allow wheelchair access and have extra grabrails in the cabins to allow greater mobility. All cabins are close to lifts.

Cabins with toilet and shower units are available on the newest of the fleet – *Normandie*, *Barfleur* and *Val de Loire*. On other vessels, there are specially constructed disabled toilets in the public areas. There is also wheelchair access to all shops, bars and restaurants on board.

All three UK ports from which Brittany Ferries sails have wheelchair access and toilet facilities for disabled people. Provision is made to assist all disabled passengers, as long as some warning is given.

'Car' discounts are available to disabled passengers on selected sailings. More information on this can be obtained from Brittany Ferries.

■ CALEDONIAN MACBRAYNE

The Ferry Terminal, Gourock PA19 1QP
☎ 01475-650100

On the newest of the larger ferries there are lifts from vehicle decks to passenger decks, toilets for disabled people and adapted eating areas. The routes that offer these facilities are:

KENNACRAIG TO ISLAY
Oban to Colonsay
Oban to Craignure (Mull)
Oban to Coll and Tiree
Oban to Castlebay (Barra) and Lochboisdale (South Uist)
Ardrossan to Brodick (Arran)
Uig (Skye) to Tarbert (Harris) and Lochmaddy (North Uist)
Ullapool to Stornaway (Lewis) – from mid-summer 1995.

Where possible, the cars of disabled motorists are placed close to lifts. Staff will be pleased to help disabled travellers. You should advise the company at the time of booking if a disabled person will be travelling.

Cars driven by or conveying a disabled person are charged at half the normal single fare for a single journey. One of the following documents must be presented to secure the discount: DDA or DDMC membership card, Tax Disc Exemption Certificate, or Gold Card (Member) of the International Disabled Travellers Club. Applications for the discount are also available for those who receive a mobility allowance (DLA). Forms are obtainable from the Caledonian MacBrayne office.

■ COLOR LINE

International Ferry Terminal, Royal Quays
North Shields, Tyne & Wear NE29 6EE
☎ 0191-296 1313

NEWCASTLE TO BERGEN
The *Viking* has two cabins equipped for disabled passengers that can take up to three people. There is a slight lip by the cabin door, but this is easily negotiated. Each cabin has been specially adapted, including the shower and WC. Cabins must be pre-booked.

A lift is available from the car deck to the passenger deck, although staff assistance may be necessary. The company should be advised at the time of booking if a disabled person is to be travelling, and efforts will be made to place a car close to the lift. All levels can be reached by lift, and all public areas, including shops and restaurants, are accessible by wheelchair. A disabled toilet is provided on the main passenger deck.

Registered disabled travellers and an accompanying passenger receive discounts of up to 50☎ on sailings at certain times of the year.

■ CONDOR LTD

Weymouth Quay, Weymouth, Dorset DT4 8DX
☎ 01305-761551

WEYMOUTH TO JERSEY AND GUERNSEY
There are no concessions for disabled travellers. Two vessels are in operation: the *Havelet* car ferry and the *Condor 10* hydrofoil. The latter has lifts to all decks.

■ HOVERSPEED

International Hoverport, Dover, Kent CT17 9TG
☎ 01304-240241

DOVER TO CALAIS (HOVERCRAFT)
At Dover a member of a party with a disabled driver or passenger must notify passenger reception or vehicle check-in on arrival. Notification may also be given prior to travel. If a wheelchair is required, this can be arranged (wheelchairs are kept in the duty office).

There are no disabled toilets in the reception area at Dover, although toilet facilities are provided in the departure lounge, to the side of the gift shop.

None of the parking spaces are marked for the specific use of disabled drivers, although three or four spaces situated against the wall just past vehicle check-in at the front of the main building are used for this purpose. Alternatively, disabled people may park in the pay-and-display car park free of charge, as long as they display the Orange Badge.

The cars of disabled people are usually loaded

onto the hovercraft first, to allow plenty of space to get in and out. Disabled passengers can also be wheeled on to the car deck separately, if required. There are a few steps from the car deck to the passenger seating area and staff will carry those with limited or no mobility to their seats. Foot passengers can also be wheeled on to the car deck and carried to their seats.

The hovercraft has no disabled toilets on board. Wheelchairs are also available at Calais.

FOLKESTONE TO BOULOGNE (SEACAT)

At Folkestone there are up to six parking spaces for disabled passengers, free for those with the Orange Badge. These spaces are in the pay-and-display car park, immediately outside the travel centre. Disabled passengers should notify a customer services assistant when checking-in.

There are disabled toilets in the departure lounge, which are at the far end of the duty-free area.

When boarding SeaCat disabled passengers have to leave the car before loading and are wheeled by a member of staff (in a Hoverspeed wheelchair) up the foot passenger ramp, onto the craft, where they can meet the rest of their party. Disabled foot passengers are wheeled on in the same way.

On board the SeaCat disabled passengers are helped to a seat by staff. If a person prefers to stay in a wheelchair, there are five places on the craft where it is safe for a passenger to remain for the duration of the crossing. Disabled toilet facilities are provided.

At Boulogne, the cabin crew prefer to take disabled passengers down to their cars before the other passengers return to their vehicles. If this is a problem, other members of the party can return to the car while the disabled passenger is wheeled from the craft with the foot passengers to a slipway at the side, to pick up the car.

The Boulogne terminal has disabled toilets on its second level, just below the duty-free area, which is itself ramped for access.

■ **IRISH FERRIES**

Passenger Office, Reliance House, Water Street
Liverpool L2 8TP   Tel: 0151-227 3131

HOLYHEAD TO DUBLIN

The *MV Isle of Innisfree* has two cabins specially adapted for disabled passengers and there are two disabled toilets.

All passenger areas are accessible to wheelchairs. Access to the outer decks involves negotiating a two-inch lip at the doors.

Three passenger lifts are accessible to wheelchairs, with one lift dedicated to disabled people

or the infirm in case of emergency.

Check-in staff should be informed by disabled drivers or those with a disabled passenger of any requirements before arriving on board. Check with the loading officer, who will try to place your car close to one of the passenger lifts.

Wheelchair access at the Dublin terminal is good, and there is access at the Holyhead terminal with notification. Staff at Holyhead can organise a van to bring disabled passengers to and from the car deck. A wheelchair is available on board the ship.

PEMBROKE TO ROSSLARE

The *MV Isle of Inishmore* has a cabin with facilities for disabled people and there is a disabled toilet. All passenger areas are on the flat, fully accessible and with wide alleyways. Lifts are available to all passenger areas, with ramps to the lift from the car deck.

■ **ISLE OF MAN STEAM PACKET COMPANY**

PO Box 5, Douglas, Isle of Man, IM99 1AF
☎ 01624-661661

You should advise the company at the time of booking if a disabled person will be travelling and reveal the nature of the disability and the kind of help that will be needed. Discounts are available to members of the DDMC and the DDA who own and drive specially adapted cars, on proof of membership.

The *MV King Orry* sails all year from Heysham to Douglas. There are toilets adapted for disabled people and a lift can be used from car deck to passenger deck. If you prefer, a wheelchair is available at the port for access via the passenger gangway.

The SeaCat *Isle of Man* serves routes from Liverpool, Fleetwood, Belfast and Dublin to Douglas, summer only. No specific facilities are provided for disabled travellers, although assistance is available from the crew.

There are disabled toilet facilities at Douglas but not at Heysham.

■ **NORTH SEA FERRIES**

King George Dock, Hedon Road, Hull
North Humberside HU9 5QA
☎ 01482-77177

HULL TO ROTTERDAM AND ZEEBRUGGE

At Hull terminal lifts and toilet facilities are provided for disabled people. Ferries are accessible by wheelchair via the passenger walkways. Disabled people travelling by car can arrange, by calling in advance, for their vehicle to be parked next to a lift on the car deck. Staff can assist passengers on and off the ship.

On board, lifts between all decks allow access for disabled passengers. Upon request, specially adapted cabins can be reserved. These cabins are situated on the same deck as the purser's office and feature extra-wide doors, allowing easy access for wheelchairs. Each cabin has an adapted toilet/shower area with supports, a drop-down seat in the shower and a low-level washbasin. All the cabins are fitted with an alarm button. There are three of these cabins on each of the vessels sailing on the Rotterdam service and one on each of the vessels on the Zeebrugge service. Passengers requiring these cabins are advised to book in advance.

Members of the DDA and the DDMC are entitled to reduced fares, whereby the car is carried at a 50% reduction (with the exception of weekends in July and August).

■ P&O EUROPEAN FERRIES

Channel House, Channel View Road, Dover Kent CT17 9TJ ☎ 01304-203388

Arrangements are made wherever necessary for all passengers with mobility or hearing problems. In all cases, however, it is important for passengers with special needs to inform the company at the earliest time that they will be travelling, ideally when a reservation is made, although measures can still be taken on the day of departure, if necessary.

Car drivers, their passengers and coach passengers with special needs will normally be boarded at a time best suited to allow unimpeded access to the ship's lifts. Once on board, disabled passengers will be assisted and guided by members of staff from the car decks to the lounges, restaurants or other public areas, using the lifts.

Drivers with disabled foot passengers using services from Dover may arrange with Dover Harbour Board for temporary parking spaces at the berth. Such spaces are limited and need to be booked in advance. Similar facilities may also be available at other ports on request.

For those travelling without a vehicle, the company has specially adapted minibuses, which can take three wheelchair-bound or five seated passengers. These are available in the UK and on the Continent to help transport wheelchair users or other passengers with disabilities. The minibuses take passengers direct to the car deck lifts. Passengers are then escorted to specified facilities on board and again assisted when required. All P&O European Ferries courtesy buses can be lowered to assist with wheelchairs (except electric wheelchairs).

All routes offer lifts from the car deck to the main passenger deck of the ship.

Special arrangements are also made for deaf passengers. For safety reasons the company would prefer them to be accompanied, but if this is not possible, passengers are asked to advise the information desk on board the vessel of their disability. One of the ship's crew can then ensure that the passenger's needs are met. In addition, a Minicom Textphone has been installed in the company's Central Reservations Office.

Registered members of the DDA or the DDMC can take their vehicle on a ferry free of charge. Passengers pay the full published standard return or standard single price.

ROUTE INFORMATION

DOVER TO CALAIS

The crossing time is 75 minutes, with up to 28 daily return sailings. Access is good, with lifts that connect all the facilities on two decks, including restaurant, cafeteria, bars, duty- and tax-free shops and bureau de change. There are also disabled toilets on board.

PORTSMOUTH TO LE HAVRE AND CHERBOURG

The crossing time is five and three-quarter hours (seven to nine hours at night), with up to three daily return sailings. All ships have wheelchair-accessible toilet facilities, lifts from car decks to upper decks, easy access through doors, reserved seats for disabled people, and accessible cabins and restaurants.

PORTSMOUTH TO BILBAO

The crossing time is 33 to 34 hours from the UK, with two return sailings weekly. Lifts are available on board to all decks. Toilets and cabins are provided for disabled passengers. The passenger terminals at Portsmouth and Bilbao have disabled toilets.

FELIXSTOWE TO ZEEBRUGGE

The crossing time is four hours, with up to eight daily return sailings. There are lifts from car decks to all other decks, and toilets and cabins adapted for disabled travellers are available on all vessels. Accessible toilet facilities are provided in the passenger terminal at Felixstowe. There is also a minibus with a ramp to take passengers with limited mobility to the ship. A wheelchair is available with steward assistance.

CAIRNRYAN TO LARNE

The crossing time is two and a quarter hours, with up to eight daily return sailings. Facilities on both ships comprise lifts to all main passenger areas, accessible toilet facilities, and wide entrance doors to the main public areas such as the restaurant, lounges and bar, which can all accommodate wheelchairs.

■ P&O SCOTTISH FERRIES

PO Box 5, Jamieson's Quay, Aberdeen AB9 8DL
☎ 01224-572615

ABERDEEN TO LERWICK
ABERDEEN TO STROMNESS
ORKNEY TO SHETLAND/SHETLAND TO ORKNEY

The *MV St Clair* and *MV St Sunniva* have lifts, ramps and cabins adapted for disabled passengers. Assistance is available from staff, and adapted toilets are provided on the *MV St Clair*. Aberdeen and Lerwick terminals have access for disabled travellers.

SCRABSTER TO STROMNESS

The *MV St Ola* has toilets adapted for disabled passengers and assistance is available from staff.

There is a 50☎ fare discount for members of the DDA or the DDMC. Members should obtain a ferry reservation form from their association secretary and forward the completed form to P&O Scottish Ferries at the above address.

■ RED FUNNEL GROUP

12 Bugle Street, Southampton
Hampshire SO14 2JY ☎ 01703-330333

SOUTHAMPTON TO EAST COWES

Two new car ferries, the *Red Falcon* and the *Red Osprey*, operate on a crossing that takes about 55 minutes.

SOUTHAMPTON TO WEST COWES

This is a high-speed passenger-only service with a journey time of around 22 minutes.

All vessels have lifts and ramps. The two car ferries have disabled toilets and access to public areas. When making a reservation, ask for assistance on the day of travel. All terminals have disabled toilets.

Special rates are available for vehicles belonging to members of the DDMC and the DDA, or any disabled drivers who can be vouched for by their local social services.

■ SALLY LINE

Argyle Centre, York Street, Ramsgate
Kent CT11 9DS ☎ 01843-595522

RAMSGATE TO DUNKERQUE

Two ferries operate on this route – the *Sally Star* and the *Sally Sky* – and the crossing takes two and a half hours.

RAMSGATE TO OOSTEND

Oostende Lines operates three ferries – the *Prins Albert*, *Reine Astrid* and *Prins Filip* – and two jetfoils on this route. The ferries make six return crossings daily, the jetfoils up to six flights daily depending on the season. The crossing takes four hours by ferry, 100 minutes by jetfoil.

The terminal at Ramsgate has disabled toilets. All ferries have disabled facilities, including toilets and lifts. Please mention at the time of booking if a passenger is disabled.

Special rates are offered to members of the DDMC. Members' cars travel free, while all adult passengers pay an extra car passenger fare which varies depending on the length of stay and the time of travel. Applicants must send a completed Sally booking form to the DDMC for a letter of approval. On receipt of this, Sally will make the requested booking.

■ SCANDINAVIAN SEAWAYS DFDS LTD

Scandinavia House, Parkeston Quay, Harwich
Essex CO12 4QG ☎ 01255-240240

HARWICH TO ESBJERG
NEWCASTLE TO ESBJERG (MARCH TO OCTOBER)
HARWICH TO GOTHENBURG
NEWCASTLE TO GOTHENBURG (JUNE TO SEPTEMBER)
HARWICH TO HAMBURG
NEWCASTLE TO HAMBURG (MARCH TO OCTOBER)

All Scandinavian Seaways passenger vessels, except for the *Winston Churchill*, are equipped with cabins for disabled travellers. These are held exclusively for disabled passengers up to one week before departure.

The *Dana Anglia* has one adapted four-berth outside cabin and two four-berth inside cabins. The doorway to the outside cabin is 96cm wide, with an 85cm opening into the bathroom. There is a toilet with chair and handrails, and a shower with chair. The doors have no steps. The inside cabins have 78cm wide doors but they do have steps. A disabled toilet and shower are in the same corridor.

The *Prince of Scandinavia* and the *Princess of Scandinavia* have two four-berth outside cabins and two two-berth inside cabins. The doors have no steps and are 77cm wide. The bathroom has a shower with chair.

The *Queen of Scandinavia* has ten two-berth outside cabins. Doors are 77cm wide, with a small step. The shower and toilet are equipped with handrails.

The *King of Scandinavia* has two four-berth inside cabins. The door to the cabins is 77cm wide, with a 72cm gap to the bathroom, plus a small step. The shower has a bench.

The *Hamburg* has two two-berth inside cabins,

with wide doors, level flooring and shower and toilet room. There are also handrails next to the washbasin and shower/toilet.

There are lifts in Gothenburg, Harwich and Esbjerg. At Harwich there is a lift operated by British Rail between the footbridge and platforms 2 and 3, and also a lift from platform 1 to the new departure terminal.

Disabled drivers should endeavour to arrive in good time for all sailings, as arrangements can then be made for the car to be located on deck near a lift. Please note that there is a high step from the car deck into the lift (except on the *Queen of Scandinavia*) and assistance will be required.

■ STENA SEALINK LINE
Charter House, Park Street, Ashford
Kent TN24 8EX ☎ 01233-647047

Stena Sealink will make arrangements for anyone who requires assistance at ports or on ships. Please let the company know in advance. Arrangements should be made through the Customer Services Department or by contacting the port of departure at least 24 hours prior to the day of travel.

STRANRAER TO LARNE
Disabled toilet facilities are provided in both the main terminal and foot passenger terminal at Stranraer. A wheelchair and courtesy coach are available, and assistance will be given to unaccompanied disabled passengers on request.

All ships on this route have lifts and disabled toilets. The *Stena Galloway* has restricted access to the lounge as lifts only go to the boat-deck level.

HOLYHEAD TO DUN LAOGHAIRE
At Holyhead there are disabled toilets on the station concourse, in the embarkation hall and in the car ferry lounge. Disabled access to ships is provided from the two multi-purpose berths.

Both ships on this route have lifts and disabled toilets. On the *Stena Hibernia* all facilities on deck are accessible, including the duty-free shop. On the *Stena Cambria*, passengers with a wheelchair are advised to inform the loading officer prior to loading in order to reserve suitable access to exits. Those in wheelchairs and not with a vehicle should board over the vehicle ramp to get to a lift. There is no suitable access from the gangway.

FISHGUARD TO ROSSLARE
Disabled toilets are located on the main station concourse at Fishguard; assistance to the concourse can be provided, with the use of a wheelchair if required. Car parking spaces are available for disabled motorists.

The ship on this route, the *Stena Felicity*, has lifts and disabled toilets. Two adapted three-berth cabins, next to lifts, are available for disabled passengers. Vehicles with disabled passengers may be loaded early and parking can often be arranged next to a lift. Assistance can be provided to passenger accommodation.

SOUTHAMPTON TO CHERBOURG
Disabled passengers must advise the travel centre of their arrival. Southampton has lifts and ample parking, with easy access to the terminal and disabled toilets. Access to the vessel for foot passengers is very limited, although assistance is given. There is access via the car deck lifts for disabled passengers travelling by car, and early access is available, with parking next to the lifts.

Cherbourg has ample parking, easy access to the terminal and toilets for disabled people.

The ship on this route, the *Stena Normandy*, has lifts and a disabled toilet. All decks are accessible by wheelchair via the lifts. Two two-berth cabins have wheelchair access and adapted toilets.

NEWHAVEN TO DIEPPE
Newhaven and Dieppe have ample parking and are easily accessed by car. Disabled parking spaces are available in the Newhaven port area. Assistance can be given to unaccompanied disabled passengers on request. Disabled toilets are provided in the Newhaven terminal.

Both ships on this route have a lift and disabled toilet. Vehicle parking can be arranged next to the lift if passengers arrive at the port one hour before sailing. The hazard lights should be flashed as the vehicle approaches the ship. Assistance will be provided on request and the *Stena Londoner* has a wheelchair available.

DOVER TO CALAIS
The Dover terminal has a lift and a disabled toilet. For sailings from the Eastern Docks, disabled passengers who arrive by train will be taken by bus from the railway station to the berth, with a stop at passport control where all passengers are required to get off the bus to show their passports before continuing on to the ship. Those with serious disabilities may find this a problem. A vehicle specially adapted for disabled passengers is available on request.

For sailings from the Western Docks, disabled passengers who arrive by car can be dropped near to passport and customs controls, from where they will proceed to the gangway. If arriving by train, the passenger travels down a short ramp at the end of the railway platform, through passport and customs controls to the gangway.

The Calais terminal has a lift and full disabled toilet facilities. The terminal building and passport controls are some way from the berths, and all movement between them must be in a vehicle. Disabled parking spaces are provided in the port area and assistance will be given to unaccompanied disabled passengers on request.

All ships on this route have lifts and adapted toilet, and offer access to most public areas.

HARWICH TO HOOK OF HOLLAND

At Harwich there are specially adapted lifts from the railway platform to the passenger hall. Disabled toilets are situated in the passenger hall and car hall reception. Those travelling to the port by train should contact British Rail about any particular requirements. Ramps are available at Parkestone Quay for safe access from the train to the platform. Car parking spaces are provided for disabled drivers.

Both ships on this route have lifts and disabled toilet facilities. Vehicles with disabled passengers may be loaded early and parking can be arranged next to a lift. Assistance can then be provided to the passenger accommodation. The *Stena Britannica* has four two-berth adapted cabins, with wheelchair access, an alarm to the information desk and a safety night light. The bathrooms are fitted with grabrails around the washbasin and toilet. This ship also has two wheelchairs for use by disabled passengers.

The *Koningen Beatrix* has two four-berth adapted cabins, with wheelchair access and bathrooms fitted with grabrails.

■ SWANSEA CORK FERRIES

Kings Dock, Swansea SA1 8RU
☎ 01792-456116

SWANSEA TO CORK

A 10-hour crossing operates up to six times a week.

Disabled travellers should detail any special requirements at the time of booking.

Two cabins can be booked in advance which are partly adapted for disabled passengers. Access to the cabin itself is not difficult, but access to the toilet and bathroom area is not level. There is a small lift from the car deck to the passenger deck, which can hold a person in a wheelchair and one companion. Disabled travellers' vehicles will, where possible, be placed close to the lift. The cafeteria and restaurant cannot be reached by lift.

Special rates apply for members of the DDA and the DDMC on proof of membership. They may travel at the foot-passenger rate, with no charge for the car. These benefits are not available in July and August.

THE CHANNEL TUNNEL

The Channel Tunnel is actually three tunnels, each 31 miles (50 km) long. Two outer tunnels carry railway tracks for the two main thoroughfares – Le Shuttle passenger and vehicle service, and the Eurostar passenger train – with a connected service passage between.

Slip-roads from the M20 (junction 11A) near Folkestone and the A16 (junction 13) near Calais take cars direct to the terminals. Passengers pay at a toll-booth, go through frontier controls and load their cars onto the special carriages. The crossing takes about 35 minutes, and the whole process, including customs, embarkation and disembarkation, should take around an hour.

Up to four trains an hour operate at peak times, with a minimum of one an hour at night. The operation is run on a strictly first-come, first-served basis. You cannot book space on a particular train, but you can pre-purchase a ticket from Le Shuttle Customer Service Centre or from travel agents.

Food, drink and duty-free goods are available at the terminals, but there are no refreshments on the train itself. Train services direct from London to Paris and Brussels, and freight services, are run by British Rail, SNCF and the Belgian SNCB, with their own trains and stations.

LE SHUTTLE

Le Shuttle Customer Service Centre, PO Box 300 Cheriton Parc, Folkestone, Kent CT19 4QD
☎ 01303-271100

This has the potential to be the simplest (if not the cheapest) way for a disabled motorist to cross the Channel as you do not have to leave your car for the duration of the trip.

Measures have been taken to make the journey as easy as possible for disabled passengers. Disabled people in wheelchairs will find good access at Folkestone and Calais terminals, and all buildings have ramps.

Disabled passengers are recommended to notify the Toll Plaza on arrival.

Car parks and waiting areas have spaces reserved for disabled motorists. Vehicles carrying disabled passengers will be identified by Le Shuttle crew and directed on to Le Shuttle on a first-on and first-off basis. The emergency crews will therefore know where the disabled travellers are so they can offer immediate assistance in the case of an emergency.

Shuttles are equipped with their own wheelchairs, which are narrower than an average wheelchair, but will speed evacuation should the need arise. There are no disabled toilets on Le Shuttle itself.

Hiring a Car at Home and Abroad

Hiring a car if you are a disabled person can be difficult and expensive. Since the last edition of *On the Move*, the options for disabled drivers to hire a car specially adapted with hand controls have narrowed. As we went to press, Hertz Car Rentals had temporarily suspended its service of hiring out hand-control cars for disabled drivers. This situation may be reviewed – contact Hertz for more information. The service detailed below, however, offers a number of hire options for disabled people.

Wheelchair Travel Ltd

1 Johnston Green, Guildford, Surrey GU2 6XS
☎ 01483-233640

The company offers the hire of vehicles to those wishing to travel in the UK or to take a vehicle abroad. Visitors of restricted mobility arriving from abroad at Heathrow, Gatwick or Stansted airports can hire self-drive wheelchair-adapted vans (Mazda, Ford and Renault) for their UK holiday. Groups from two to eight people where at least one is a wheelchair user can use the service.

A self-drive van and car hire service is also offered within the UK for disabled people. In total there are 13 minibuses, three 'Chairman' vehicles and two hand-controlled cars (a Ford Sierra and Vauxhall Carlton, both automatic estates) available for hire. At the time of writing, a three-door hatchback car was due to be added to the fleet.

Vehicles may be delivered to and collected from a client's home, or an airport, depending on distance, availability and date and time, for an extra cost. People may take vehicles to Europe, but must first obtain extra comprehensive insurance and breakdown cover.

For more details, contact Trevor Pollitt at the above address.

UK Airports

It is very important that your travel agent and airline should be made aware at the time of booking of your particular requirements and what assistance you will need, particularly through the terminal buildings, and on boarding and leaving the aircraft. It is essential that you give clear and detailed information about your disability and its effects. You should check whether your airline makes any charge for special assistance, such as the use of a wheelchair or ambulance en route to your destination. It is worth finding out all the details (such as methods of transport to and from the aircraft) and comparing the arrangements of the various airlines and tour operators before making your booking. You may find that you need a form completed by your doctor, stating the nature of your disability and confirming that you are able to travel.

■ HEATHROW

☎ 0181-759 4321

Heathrow is 15 miles west of central London, just south-east of the M4/M25 junction. Terminals 1 to 3 are in the Central Terminal area, which is reached through the tunnel from the M4 spur and from the A4. Terminal 4 is south of terminals 1 to 3 and just off the A30. It is reached from London via the M4, junction 3. Motorists coming east on the M4 should change to the M25 at junction 4B/15. From the M25, leave at junction 14 for Terminal 4.

The London Underground Piccadilly Line 0181-222 1234-stops first at Terminal 4, and then at the central station, serving terminals 1 to 3. All terminals and the central bus station are connected to central London by the wheelchair-accessible Airbus (0181-897 3305).

There are wide parking spaces reserved for Orange Badge holders in all the short-stay car parks, but these do cost £30 a day. More economical is the Flightpath long-term car park at £7 per day. An accessible coach is available on request. There are also business and valet parking services. Telephone 0181-745 7160 for details of all car parking arrangements. For security reasons, the Orange Badge scheme does not apply at Heathrow Airport and vehicles should never be left unattended.

All passenger areas of the airport are accessible via lifts, ramps or walkways. There are accessible, usually unisex, toilets in all terminals. Eating and shopping areas can be found in most parts of the airport, but departing passengers will find more extensive facilities after they have passed through passport control.

For further information about travelling through Heathrow Airport, call Heathrow Travel-

Care, the independent airport social work agency, on 0181-745 7495 (Minicom: 0181-745 7565 - (Monday to Saturday 9.30am to 4.30pm).

■ GATWICK

☎ 01293-535353/531299 (flight enquiries)

Gatwick Airport has two terminals, North and South, both reached via a spur road from junction 9 of the M23. An automatic transit system, which runs continuously, links the two terminals.

Details of which flights use which terminal change frequently, so check on your ticket for flight details or consult the airline. There is also a 24-hour information line on 01293-567675.

Each terminal has its own short-term, multi-storey car park and a long-term car park with a free bus shuttle service to the terminal. All are clearly signposted on the access roads.

For full details of the types of assistance available for disabled travellers, contact your airline's handling agent:

British Airways Gatwick Handling	01293-666291
North Terminal	01293-507147
South Terminal	01293-502337
Servisair	01293-507320

Courtesy telephones linked to the handling agents are available at entrances to both terminals and in parking bays for disabled people. If advised in advance, your handling agent will also provide help on the arrival of your return flight.

The British Rail station is linked to South Terminal. Passengers for North Terminal should follow the signs from the station platforms and take the transit. Disabled travellers can arrange with their local station manager for station staff at Gatwick to assist on their arrival. For further advice call 01293-524167, the disabled person's rail travel line.

Taxis accessible to wheelchair-bound passengers are available. All cars have meters and may be booked in advance from Gatwick Airport Cars, ☎ 01293-562291.

Parking

Bays for Orange Badge holders are reserved in the multi-storey car parks of both terminals, for short- or long-stay parking, giving level access to the terminals. For any stay over eight hours the long-term rate applies. Courtesy telephones are available. For further details of short-term multi-storey car parks call Parking Services Granada on 01293-502390.

Bays are also reserved in the long-term car parks of both terminals. For further details call:

North terminal (Flightpath) 01293-502748
South terminal (Flightpath) 01293-502679
North terminal (Parking Express) 01293-502357
South terminal (Parking Express) 01293-502896
Gatwick Airport Ltd Car Park
  Operations 01293-503897

If you are leaving a vehicle unattended to set down or pick up passengers, use the reserved bays in the multi-storey car parks.

Flight departures

Passengers with difficulty in seeing monitors and signs should advise their handling agent at check-in.

Passenger services

Passengers with a 'T' position on their hearing aid can receive announcements relayed over the public address system when standing close to the induction loops. These are indicated by the 'ear' symbol. Some staff are proficient in sign language and can be identified by a lapel badge bearing the 'ear' symbol. The airport information desk in North Terminal is equipped with an induction loop to assist those with hearing aids. South Terminal's information desk has a Minicom Supertel telephone – 01293-513179 – for the deaf and hard of hearing.

For passengers with wheelchairs, there are ramps and/or lifts at every change of level. Reserved seating is provided at the check-in areas. Telephones are at accessible heights for wheelchair users, and disabled toilets are available throughout both terminals.

■ BIRMINGHAM

☎ 0121-767 5511

The airport is south-east of Birmingham, next to the National Exhibition Centre and just off the A45 and M42.

If you are driving and intend to stay for less than four hours, spaces are reserved in the short-stay car park close to the terminal. For longer periods, spaces on the ground floor of the multi-storey car park are available, next to the main terminal, with access via a covered walkway. If you need assistance, press the control barrier 'assistance' button at any of the car park entrances and exits. The car parking charges are at the normal commercial rates and parking outside the terminal buildings is limited to a waiting time of five minutes. For more information on car parking, contact National Car Parks on 0121-767 7861.

Help with baggage and check-in is available by contacting the terminal information desk on arrival. All passenger facilities are on the ground and first floors and are easily accessible, with signposted lifts. Wheelchairs are available on request. A Minicom system – 0121-782 0158 – is installed at the main terminal information desk.

Toilets are provided for disabled people throughout the terminal buildings and also in the multi-storey car park (keys can be obtained from the car park control room).

Airlines make their own arrangements for the care of disabled travellers and help will be given to get you to and from the aircraft, using either a wheelchair or a purpose-built lift vehicle.

■ BRISTOL

☎ 0275-474444

The airport is eight miles south of the city on the A38. From the south, leave the M5 at junction 22 and follow the A38 towards Bristol. From the north, leave the M5 at junction 18 and follow signs to the airport.

Designated spaces are available for disabled passengers in the short- and long-term car parks. Wheelchairs, low-level telephones, a lift, toilet facilities and a passenger transfer vehicle are all provided for disabled travellers.

■ CARDIFF

☎ 01446-711111

The airport is 10 miles west of Cardiff, near Barry, and is reached from the M4. Leave at junction 33 and follow signs for the airport.

The main car park is close to the terminal. A limited number of special car parking spaces are available on request from airport security – call 01446-711111. Wheelchairs are available free of charge on request from your airline or its handling agent. There are lifts to all floors of the terminal, and accessible toilets and telephones. Ramps and automatic doors make for easy access to the terminal building. A custom-built wheelchair lift enables disabled travellers to use their own wheelchairs when going up or down two flights of stairs on the International Pier. Also, the three airbridges serving the larger aircraft allow disabled people to board direct from the terminal building.

■ EAST MIDLANDS

☎ 01332-810621

The airport is situated at Castle Donington, near Derby, two miles from junction 24 of the M1. Access from the West Midlands is via the A453 and A42/M42.

When you arrive at the airport you can drive up to the front of the terminal, where there are ramps, lowered paving, special parking places and a Servicecall sensor. When activated, this will alert a member of the Customer Services team, who will come to your vehicle to give assistance. NCP has set aside parking spaces for disabled drivers close to the terminal. If you want to reserve a parking space, call 01332-810621.

Once in the terminal, if you want help, or have any queries, ask one of the Customer Services team at the information desk. Wheelchairs can be provided and an amplified telephone facility is available. Information is also offered in braille.

Adapted toilets and lowered telephones are available for disabled passengers. Seating in the restaurant is movable and a lift can take passengers to the first-floor refreshment and viewing area.

If you have problems negotiating aircraft steps, a carrychair or hydraulically operated ambulift may be used to help you to board.

■ LEEDS/BRADFORD

☎ 0113-250 9696

The airport is eight miles north-west of Leeds, six miles north of Bradford, and eleven miles south of Harrogate. It is passed by the A658, which runs north to Harrogate and south to Bradford. From Leeds, take the A65 and then the A658 north at Rawdon. From the M1, take junction 42 to the M62, and then junction 28, following the signs for the airport.

Wheelchairs, ramps, toilet facilities, ambulift transport and special car parking arrangements are all available for disabled travellers.

If you need further help, ask at the airport information desk, opposite the main doors on the ground floor of the terminal building.

■ LIVERPOOL

☎ 0151-486 8877

The airport, at Speke, is easily accessible from the M6, M62 and M56. From the north, leave the M6 at junction 21A, take the M62 west to junction 4, then the A5058/A562 ring road north to the junction with the A561. Follow signs to Airport Terminal, situated off Speke Hall Avenue. From the south, leave the M6 at junction 20, take the M56 to junction 12, then the A557/A562 west via Runcorn Bridge to merge with the A561.

A pick-up and set-down area is available immediately outside the terminal entrance. Disabled parking spaces are reserved at the front of the car park.

Most passenger facilities are based on the ground floor of the terminal. The airport buffet, bar and seating area on the first floor are accessible via a lift to the left of the information desk. Disabled toilets are available on both the ground and first floors of the terminal. Access to disabled toilets is also available from the departure and arrivals halls. Low-level telephones can be found in the main concourse, close to the arrival and departures areas.

Wheelchairs are available at all times and should be requested when booking your flight. If you need a wheelchair when in the terminal, ask at the information desk. The Airport Fire Service assists wheelchair passengers to pass through the airport for departures and arrivals.

The airport has its own ambulift to automatically lift disabled passengers and wheelchairs to aircraft door sill height.

■ LUTON

☎ 01582-405100

The airport is south-east of the town centre, just off the A505. From the M1, leave at junction 10, whether you are coming from the north or south. From the A1(M) northbound leave at junction 8 via the A602 to the A505, southbound via junction 9.

All facilities at Luton Airport are on ground level and designed with disabled travellers in mind. Ramps and automatic doors ensure easy access to the terminal, and there are disabled toilets in the check-in area, departure lounge and baggage reclaim hall.

Wheelchairs are available on request at airline check-in desks, and specialised equipment can be provided to help disabled passengers board the aircraft. This should be requested in advance through your travel agent, tour operator or airline. All major airlines operating from Luton will transport your wheelchair free of charge.

A medical centre, staffed by qualified nurses, is located in the check-in concourse.

A dedicated disabled car park, at the front of the terminal, offers special help. Disabled drivers should write or telephone – 01582-395273 – in advance to the Car Park Controller. The normal car parking charge will apply. Disabled visitors to the airport may use the spectators' car park free of charge (for stays of up to 12 hours) as long as they have an Orange Badge.

Further information and advice for disabled travellers can be obtained by calling the airport information desk on 01582-405100. For any special needs that you may require contact:

Reed Airport Services, Eaton House, Proctor Way, London Luton Airport, Luton, Bedfordshire LU2 9LY ☎ 01582-421200

■ MANCHESTER

☎ 0161-489 3000

The airport is about 10 miles south of Manchester city centre and is reached via a spur from the M56. There are two terminals: Terminal A, which handles domestic flights, and Terminal B,

for all internation

Special parking parks at both ter Car Parks Depa

Adapted toile both terminals vehicles – aml vided to conv terminal bui are provided the level changes.

■ NEWCASTLE

☎ 0191-286 0966

The airport is at Woolsington, six miles north-east of the city. It is reached via the Central Motorway East and the A696.

There are extra-wide disabled car parking bays in a reserved area. At the front of the terminal building is a disabled help line so you can summon porter assistance without leaving your car. The entrance doors have a slow revolving mode for wheelchair users.

Disabled toilets with an alarm button are provided. There is an induction loop system in various parts of the terminal building, and at the airport information desk a Minicom telephone is installed. Payphones have an induction coupler and there are assistance telephones in the duty-free lounge. Some of the airport staff are also trained to communicate in sign language.

A lift allows access to the higher level of the terminal building, and a cabin lift allows disabled passengers to board without negotiating the aircraft steps.

■ STANSTED

☎ 01279-680500

Stansted lies north of London, just off junction 8 of the M11. The access road leaves the north-east corner of the junction roundabout. There is a rail link to London (Liverpool Street), with the station incorporated into the terminal building.

The terminal has been designed with the help of groups representing disabled people. Responsibility for looking after disabled travellers at Stansted is shared between the airlines and Stansted Airport Limited. However, from the time you check in with your airline until you collect your baggage and have cleared customs at your destination, the airline or its handling agent should provide assistance.

If you are arriving by train, your local station manager will inform the staff at Stansted Airport station of your arrival time and make arrangements for you to be met. The station is beneath the terminal forecourt, with access to and from the terminal by lift and ramp.

in front of the terminal
...ss to and from the terminal

...ing, there are reserved parking
...ge Badge holders in the short-stay
...ose to the terminal entrances, with
... lift or ramp. For assistance, you can use
...ormation telephones provided next to the
...ing bays.
...f you use the long-term car park, the operator
will provide help to get you to the terminal.
There are frequent courtesy buses and all are
fitted with wheelchair lifts and tracking.

If you are brought to the airport by car, your driver can stop briefly on the forecourt in front of the departures concourse. The car must not be unattended but parked in the short-stay car park. For help from the forecourt, you can use the information telephones next to the terminal entrances. If you are being collected by car from your flight, the car must be parked in the short-stay car park next to the terminal. If you are a wheelchair user and need special parking assistance contact:

Ground Transportation, Enterprise House Stansted Airport, Stansted, Essex CM24 1QW

Each airline or its handling agent will arrange for someone to help you when you arrive at one of the three terminals, as long as they know in advance how and when you are travelling.

Special facilities inside the terminal for wheelchair users are indicated by signs displaying the wheelchair symbol. Each check-in has two low-level desks, designed specifically for wheelchair users. Disabled toilets and accessible telephones are located throughout the terminal and in the Satellite.

In the terminal and Satellite 2 all passenger facilities are on a single level, but in Satellite 1 lifts have to be used between the Tracked Transit System and the gates. Your airline or handling agent will help.

If you have a hearing aid with a 'T' position you can receive enhanced announcements relayed over the public address system when standing close to the induction loops indicated with the 'ear' symbol.

Those with impaired sight and who cannot read the flight displays should advise their handling agent when checking in, or seek help from the airport information desk. You can contact the airport information desk in advance by calling 01279-662379 or 662520.

Special arrangements have been made for the most severely disabled travellers to enter and leave the aircraft apron via the ground floor of the terminal. Sometimes passengers can be wheeled direct to or from the aircraft in their wheelchairs, while at other times a vehicle may have to be used. Practices will vary depending on the airline.

■ ABERDEEN

☎ 01224-722331

The airport is situated at Dyce, seven miles north-east of Aberdeen on the A96. The main car park (Euro Car Parks) is next to the terminal. The terminal is accessed from road level, with dropped kerbs to the entrance. The same level is maintained through to the aircraft steps. Wheelchair accessible telephones, toilets and catering units are available throughout the terminal.

A telephone link is available in the car park, which allows disabled passengers to request assistance from a member of staff in the terminal building, particularly those in wheelchairs who need help in carrying luggage. If a car or taxi is meeting you on arrival, it is best to use the airport information desk as a meeting point. If you need any assistance, speak to a uniformed member of staff or contact the Airport Duty Manager on ext 5060.

■ EDINBURGH

☎ 0131-333 1000

The airport is six miles west of the city, just off the A8. The access road leaves the A8 east of the Royal Highland Showground, midway between Maybury and Newbridge roundabouts.

The terminal is accessed from road level by dropped kerb on to the pavement and through automatic double doors. A telephone link is available in the car park, which allows disabled passengers to request assistance from a member of staff in the terminal building, particularly those in wheelchairs who need help in carrying luggage.

A restaurant and bar are easily accessible at ground-floor level, and a lift will get you to the café on the first floor. Wheelchair accessible telephones and toilets are available throughout the terminal.

If a car or taxi is meeting you on arrival, it is best to use the airport information desk as a meeting point. If you need any assistance, speak to a uniformed member of staff or contact the Airport Duty Manager on ext 3323.

■ GLASGOW

☎ 0141-887 1111

The airport (Abbotsinch) is near Renfrew, south-west of the city. The entrance road leads off junction 28 of the M8, about 20 minutes from the centre of Glasgow. Several car parks are close to the terminal and are well signposted.

Access to the terminal is at pavement level with dropped kerbs through automatic doors. There are lifts from the ground floor to the first and second floors, as well as one on the International Pier, which gives wheelchair access to the duty-free shop.

A Minicom telephone – 0141-848 4848 – is available, as are wheelchair accessible toilets, telephones and refreshment places throughout the terminal. If you need any assistance, speak to a uniformed member of staff or contact the Airport Duty Manager on ext 4510.

■ **BELFAST INTERNATIONAL**
☎ 01849-422888

Belfast Airport is 13 miles north-west of the city and is reached from the A52 and A26 or the M2 and A26.

Wheelchairs are available on request from your airline or from Airport Services staff. Where possible, this should be arranged in advance of your trip, preferably with your airline at the time of booking. Airport Services staff will look after you from the time you arrive at the entrance hall of the terminal until you check in. After that your airline will ensure that you are taken to the correct departure lounge and will help you when boarding.

Accessible toilets and telephones are available in the terminal, and two desktop Minicom units are installed to assist deaf callers with their enquiries. There are lifts between the arrivals and departures floors.

Special parking facilities are available for disabled drivers in the airport car parks.

RECIPROCAL PARKING ARRANGEMENTS IN EUROPE FOR ORANGE BADGE HOLDERS

A system of reciprocal arrangements exists within Europe under which disabled visitors from the participating countries can take advantage of the concessions provided in the host country by displaying the badge issued under their own national scheme.

In some countries responsibility for introducing the concessions rests with individual local authorities. Concessions may, therefore, not be generally available. In such cases badge holders should enquire locally, as they should whenever they are in any doubt as to their entitlement.

As in the UK, the arrangements apply only to badge holders themselves, and not able-bodied friends or relatives. Able-bodied people who seek to take advantage of the concessions in Europe by wrongfully displaying an orange badge will be liable to whatever penalties apply for unlawful parking in the country in question.

EUROPEAN SIGNS

The basic signs are the red and blue roundels, which are also used on the signs in the UK.

In some spaces waiting is permitted on opposite sides of the road alternately. If the change is made daily these signs are used.

C18
No waiting

C19
No stopping

C20A
Waiting Prohibited
on odd
number dates

C20B
Waiting Prohibited
on even
number dates

If a different form of alternation applies, the numerals I and II are replaced by the appropriate dates – e.g. 1 - 15 and 16 - 31 if the change is made on the 1st and 16th day of the month.

The sign used to indicate a parking place is the same white and blue sign as is found in the UK. At the entry to a zone in which all parking is subject to a time limit, the C18 or E23 symbols may be shown on a rectangular backing. the sign is then referred to as C21. There may be a symbol on the lower part of the panel to show the system of limitation in force.

The roadmarkings used with the upright signs differ from those in the UK. A continuous yellow line on the kerb or the edge of the carriageway indicates that waiting and loading are restricted. A broken yellow line on the kerb or carriageway or a zigzag line on the carriageway indicate waiting restrictions. Parking-place markings may be white or blue.

■ **AUSTRIA**

The Austrian scheme of parking concessions allows badge holders to park without time limit where sign C18 is used, and they may stop (even if double-parked) where sign C19 is displayed. Sign C19 is used also to indicate that badge holders may park in a pedestrian zone when loading and unloading are permitted.

Parking is also allowed without time limit at parking areas normally reserved for short-term parking.

Special parking places for disabled people's vehicles may also be set aside near locations such as hospitals and public service facilities for the care of people with disabilities.

■ BELGIUM

Badge holders may park without time limit where others may park only for a limited time.

Reserved parking places are provided for disabled people's vehicles. (These are indicated by sign E23, with an additional sign showing the disabled person symbol.)

Badge holders are exempt from paying at parking meters where local regulations explicitly provide for such payment.

■ DENMARK

Parking for up to 15 minutes is allowed for badge holders:
- where only loading and unloading are permitted
- where parking is prohibited, and
- on those parts of pedestrian areas where delivery vehicles are allowed.

Parking for up to one hour is allowed where 15 or 30 minutes parking is permitted.

Unlimited parking is allowed:
- where one, two or three hours parking is permitted (the time of arrival must be shown on a parking disc)
- at parking meters or car parks with coin machines so long as the maximum amount is paid on arrival and the parking disc is set at the arrival time.

■ FINLAND

In order to obtain the parking concessions, a disabled visitor has to acquire a Finnish parking permit from the local police by presenting to them a parking permit granted in his or her own country. This permit must be displayed in a visible place inside the windscreen when the vehicle is parked.

Any badge holder granted a permit by the police is entitled to park, free of charge, in a parking space where parking is subject to payment and in an area where parking is prohibited by road signs (C18 and C21), provided that other traffic regulations do not hinder it. These concessions apply only to the vehicle indicated on the permit.

■ FRANCE

Responsibility for parking concessions in urban areas rests with the local (rather than regional or national) authorities.

Apart from reserved parking spaces for people with disabilities (indicated by the international symbol), there are no general street parking concessions. However, the police are required to show consideration towards parking by identifiable disabled people's vehicles – providing they are not causing an obstruction.

There are a number of local concessions, which vary from place to place. In Paris, for example, disabled people are exempt from street parking charges and benefit from a 75% reduction in car parking charges.

■ GERMANY

Badge holders are allowed to park:
- for a maximum of three hours where sign C18 is displayed (the time of arrival must be shown on a parking disc)
- beyond the permitted time in an area covered by sign C21
- beyond the permitted time where sign E23 is displayed, or at pavement parking areas for which the parking period is restricted by a supplementary sign
- during the permitted periods for loading and unloading in pedestrian zones
- without charge or time limit at parking meters, unless other parking facilities are available within a reasonable distance.

Reserved parking spaces for people with disabilities are also provided.

■ GUERNSEY

Badge holders may occupy short-term approved parking spaces, usually for a half-hour or one-hour limitation, for up to two hours.

■ ITALY

Responsibility for the concessions rests with local authorities. In general, public transport is given priority in town centres, and private cars may be banned. The authorities, however, are required to take special measures to allow badge holders to take their vehicles into social, cultural and recreational activity areas as well as to their work places.

Reserved parking bays are provided, indicated by signs with the international symbol. (A small number of spaces are reserved for particular vehicles, and in such cases the sign will show the appropriate registration number.)

■ JERSEY

Badge holders may park:
- in any disabled parking areas specifically set aside on-street up to a maximum of four hours (A parking disc must be displayed)
- for unlimited time in specially reserved spaces in certain meter parking areas and certain multi-storey car parks
- for up to a maximum of two hours using a normal parking disc on any disc parking area in the 20-minute (yellow) or 1-hour (red) zones.

Full details of the scheme operating in Jersey are contained in a leaflet available from The Town Hall, St Helier, Jersey.

■ LUXEMBOURG

In most urban areas reserved parking places are provided and indicated by signs C18 or E23 with the international symbol. Generally, badge holders may not exceed the parking time limit.

■ NETHERLANDS

Badge holders are entitled to the following concessions:
- the use of a car park set aside for disabled people. (The place must be signed and there is no time limit)
- indefinite parking in blue zones
- indefinite parking at places marked with the E23 sign in conjunction with an additional panel stating parking time
- parking at places marked with signs C18 and C20A or B for a maximum of two hours. A disabled person's parking disc must be used.

This concession does not apply where other parking facilities exist within a reasonable distance.

■ NORTHERN IRELAND

The Orange Badge Scheme in Northern Ireland provides concessions identical to those available in the rest of the UK. However, in many towns there are security-control zones where vehicles cannot be left unattended. For full information consult the police.

■ PORTUGAL

Parking places are reserved for badge holders' vehicles. These are indicated by signs with the international symbol.

Badge holders are not allowed to park in places where parking is prohibited by a general regulation or a specific sign.

■ SWEDEN

Badge holders are allowed to park:
- for three hours where parking is banned or allowed only for a shorter period due to local regulations
- for a period of 24 hours where a time limit of three hours or more is in force
- in reserved parking spaces indicated by signs with the international symbol.

In general, parking charges must be paid, although there might be local exemptions. The local police can give further information.

■ SWITZERLAND

Badge holders are allowed to park:
- for four hours where parking is restricted to 20 minutes or more, or within blue zones
- for two hours where parking is prohibited altogether, or within red zones.

These concessions apply only if there is no public or private parking place with unlimited parking time available to the public in the immediate vicinity. Badge holders are required also to observe any special police restrictions, regulations in private parking areas, traffic rights of way, parking time limits shorter than 20 minutes and no-stopping zones.

Concessions at parking meters vary from town to town. Badge holders should enquire at a local police station to see whether concessions are available.

■ OTHER COUNTRIES

There are no formal reciprocal arrangements with Norway, but it is understood that badge holders are allowed to park in ordinary meter or paying parking spaces free of charge and without time limit. They may also use spaces marked with the sign 'reserved parking for disabled persons entitled to parking concessions', but these spaces may be subject to limitation.

In Spain most large towns and cities operate their own schemes, but types of badges used and the concessions provided are not standardised. It is understood, however, that consideration would be shown to badge holders from other countries.

There are no reciprocal arrangements with the Republic of Ireland, but it is understood that it is national policy on the part of the enforcement authorities to make parking possible for all disabled drivers.

Useful Addresses

Organisations

ACCESS COMMITTEE FOR ENGLAND
c/o RADAR, 12 City Forum, 250 City Road, London EC1V 8AF ☎ 0171-250 3222, Minicom: 0171-250 4119
Independent policy advisory committee on the accessibility of the built environment.

AGE CONCERN
Astral House, 1268 London Road, London SW16 4ER ☎ 0181-679 8000

AIR TRANSPORT USERS COMMITTEE
5th Floor, Kingsway House, 103 Kingsway, London WC2B 6QX ☎ 0171-242 3882

ARTHRITIS CARE
18 Stephenson Way, London NW1 2HD
☎ 0171-916 1500/0800 289170 (helpline Monday to Friday 12 noon to 4pm)

ASSOCIATION FOR SPINA BIFIDA AND HYDROCEPHALUS
ASBAH House, 42 Park Road, PeterborougPE1 2UQ. ☎ 01733-555988

THE ASSOCIATION OF BRITISH INSURERS
51 Gresham Street, London EC2V 7HQ
☎ 0171-600 3333 Fax: 0171-696 8999

BRITISH LIMBLESS EX-SERVICEMEN'S ASSOCIATION
Frankland Moore House, 185/187 High Road, Chadwell Heath, Essex RM6 6NA
☎ 0181-590 1124

BRITISH POLIO FELLOWSHIP
Bell Close, West End Road, Ruislip, Middlesex HA4 6LP ☎ 01895-675515

BRITISH RED CROSS SOCIETY
9 Grosvenor Crescent, London SW1X 7EJ
☎ 0171-235 5454

BRITISH SCHOOL OF MOTORING
Disability Training Centre, 81/87 Hartfield Road, London SW19 3TJ ☎ 0181-540 8262

CENTRE FOR ACCESSIBLE ENVIRONMENTS
Nutmeg House, 60 Gainsford Street London, SE1 2NY ☎ 0171-357 8182 Fax: 0171-357 8183

A chair that could change your life

If arthritis, a bad back or other disability makes sitting an uncomfortable experience for you, then A.J. Way have a chair that could change your life. It is the result of eight years co-operation between medical experts and the craftsmen who designed and developed it. It is called the Multi-Chair.

Each chair is made to suit your own personal requirements. They are motorised to help you sit, recline and stand up again – all at the touch of a button. Find out more about these remarkable chairs.

A.J. Way & Co Ltd

Unit 2, Sunters End, Hillbottom Road, Sands Industrial Estate, High Wycombe, Bucks HP14 4HZ.
Tel: (01494) 471821 Fax: (01494) 450597

Useful Addresses

COMMITTEE ON ACCESS AND MOBILITY FOR SCOTLAND
c/o Disability Scotland, Princes House,
5 Shandwick Place, Edinburgh EH2 4RG
☎ 0131-229 8632

CONSUMER INSURANCE SERVICES
2 Osborne Court, High Street South, Olney,
Bucks MK46 4AA ☎ **01234-713535**
Fax: 01234-241191

CONSUMERS' ASSOCIATION
2 Marylebone Road, London NW1 4DF
☎ 0171-830 6000

DISABILITY ALLIANCE EDUCATIONAL AND RESEARCH ASSOCIATION
1st Floor East, Universal House,
88/94 Wentworth Street, London E1 7SA
☎ 0171-247 8776/247 8767 (rights advice line)

DISABILITY INFORMATION TRUST
Mary Marlborough Centre, Nuffield Orthopaedic Centre, Headington, Oxford OX3 7LD
℅ 01865-227592
With the help of disabled people, the Trust assesses and tests a wide range of equipment and publishes the results of these findings in handbooks – see 'Useful publications'.

DISABILITY LAW SERVICE
Room 241, 2nd Floor, 49-51 Bedford Row
London WC1R 4LR ☎ 0171-831 8031
Free legal advice for disabled people.

DISABILITY SCOTLAND
Princes House, 5 Shandwick Place, Edinburgh
EH2 4RG ☎/Minicom: 0131-229 8632

DISABLED LIVING FOUNDATION
380/384 Harrow Road, London W9 2HU
☎ 0171-289 6111
Practical information on any aspect of disability.

DSS (WAR PENSIONS AGENCIES)
Norcroft, Blackpool, Lancashire, SY5 3WP
☎ 01253-858858 *(for WPMS enquiries)* 0345 123456 *(for DLA enquiries)*

GREATER LONDON ASSOCIATION OF DISABLED PEOPLE
336 Brixton Road, London SW9 7AA
☎ **0171-274 0107 Fax: 0171-274 7840**

HAEMOPHILIA SOCIETY,
123 Westminster Bridge Road, London SE1 7HR
☎ 0171-928 2020

HEARING CONCERN
7/11 Armstrong Road, Acton, London W3 7JL
☎/Minicom: 0181-742 9151

INDEPENDENCE
92 Mychett Road, Camberley SU16 6BT
☎ 01252-513156

THE INSTITUTE OF ADVANCED MOTORISTS
IAM House, 359/365 Chiswick High Road,
London W4 4HS ☎ 0181-994 4403,

THE LIMBLESS ASSOCIATION
31 The Mall, Ealing Broadway, London W5 2PX
☎ **0181-579 1758**

JOINT COMMITTEE ON MOBILITY FOR DISABLED PEOPLE
Woodcliff House, 51A Cliff Road,
Weston-super-Mare, Avon BS22 9SE
Contact: Tim Shapley (Sec.) ☎ **01934-642313**

MEDICAL COMMISSION ON ACCIDENT PREVENTION
35/43 Lincolns Inn Fields, London WC2A 3PN
☎ 0171-242 3176

S.A.M STILL ABLE MOTORSPORT
RALLY SCHOOL FOR THE DISABLED

Still Able Motorsport are offering a unique opportunity for the disabled driver to experience for themselves the thrills of driving an

ASTRA GTE 16V RALLY CAR

on both loose and tarmac stages
agaist the clock.
Utilising a disused airfield in Central Yorkshire the one day course will allow you to learn the skills of rally driving using our cars which are fully converted to be driven with hand controls.

For more information contact:
Dave Hawkins,
Still Able Motorsport,
Wood Farm, East Bank Road,
Sunk Island, Hull HU12 0QP
Telephone: 0964 631036

S.A.M. supports STEPPING STONES. The Yorkshire Spinal Injury Centre Appeal

MOBILITY ADVICE LINE
PO Box 1551, King's Norton, Birmingham B38 8AF ☎ 0121-459 3268
A free information service, providing advice on cars, car conversions, driving instructors, adaptations, hand controls, garages, Motability and other motoring needs, such as accessories, signs and stickers.

MOBILITY UNIT
Room S10/20, Department of Transport,
2 Marsham Street, London SW1P 3EB
☎ 0171-276 5257

THE MULTIPLE SCLEROSIS SOCIETY OF GREAT BRITAIN AND NORTHERN IRELAND
25 Effie Road, London SW6 1EE
☎ 0171-736 6267/371 8000 (helpline)

MUSCULAR DYSTROPHY GROUP OF GREAT BRITAIN
7/11 Prescott Place, London SW4 6BS
☎ 0171-720 8055

NATIONAL ASSOCIATION OF VOLUNTEER BUREAUX
St Peters College, College Road, Saltley,
Birmingham B8 3TE ☎ 0121-327 0265

NATIONAL COUNCIL FOR VOLUNTARY ORGANISATIONS
Regents Wharf, 8 All Saints Street, London N1 9RL, Contact: James Ralton ☎ 0171-713 6161
Fax: 0171-713 6300
'Umbrella' group for voluntary organisations in England.

NETWORK FOR THE HANDICAPPED
Room 240, 2nd floor, 49-51 Bedford Row,
London WC1R 4LR ☎ 0171-831 8031/7740
Fax: 0171-831 5582

PARKINSON'S DISEASE SOCIETY
22 Upper Woburn Place, London WC1H 0RA
☎ 0171-383 3513 Fax: 0171-383 5754

QUEEN ELIZABETH'S FOUNDATION FOR DISABLED PEOPLE
Leatherhead Court, Woodlands Road,
Leatherhead, Surrey KT22 0BN ☎ 01372-842204

RALLY SCHOOL FOR THE DISABLED
Woodfarm, East Bank Sunk Island, Hull HU12
☎ 0964 631036

GO UPSTAIRS, DOWNSTAIRS WHEREVER YOU WANT TO GO!

THE MOBILITY 2000 POWER WHEELCHAIR DOES THIS - AND MORE

- YES our wheelchair goes up and down steps and stairs
- HIGHLY manoeuvrable at home and at work or at school
- SMOOTHLY and safely up and down kerbs to 8 and a half inches high
- COPES easily with hills and cambers, fast on the flat
- ADJUSTABLE HEIGHT reach up or reach down to the ground
- CHILD/SMALL person seating now available

To gain your greater freedom and independence contact us now for more information.

Name _____

Address _____

_____

Postcode _____ Tel No. _____

Mobility 2000 (Telford) Ltd., Telford Industrial Centre,
Stafford Park 4, Telford, Shropshire, TF3 3BA. Telephone: 01952 290180

The TRIMCHAIR™
Expanding Fitness Horizons

NOW WITH PASSIVE LEG EXERCISER AND NEW EASY ACCESS FEATURE

The TRIMCHAIR™ is a mobile exercise system which introduces progressive resistance workouts to exercise the upper and lower body.

Innovative and easy-to-use, the TRIMCHAIR™ is ideal for hospitals, sports medicine facilities, rehabilitation centres, retirement homes and private residences.

V&A Marketing
Telephone: 01222 664564
231a Cathedral Road, Cardiff
Developed & Manufactured in the UK

The Finest 🇬🇧 British Specialised Weatherwear for Wheelchair and Scooter Travellers

From the European Brand Leaders

ADULT COSYSIT

RIDING CAPE

Just Two of our many famous products

SIMPLANTEX *Health Care Limited*

SEND TODAY FOR FREE FULL COLOUR CATALOGUE

SIMPLANTEX HEALTH CARE LIMITED Dept.RAC 1, Healthcare House, 55, Willowfield Road
Eastbourne, East Sussex BN22 8AP Tel: (01323) 411618 Fax: (01323) 412781

Useful Addresses

THE RESTRICTED GROWTH ASSOCIATION
103 St Thomas Avenue, West Town, Hayling Island, Hampshire PO11 0EU ☎ 0889-576571

ROYAL NATIONAL INSTITUTE FOR THE BLIND
224 Great Portland Street, London W1N 6AA
☎ 0171-388 1266

ROYAL NATIONAL INSTITUTE FOR THE DEAF
105 Gower Street, London WC1E 6AH ☎ 0171-387 8033/0800 413114 (information line)

ST JOHN AMBULANCE BRIGADE
1 Grosvenor Crescent, London SW1X 7EF
☎ 0171-235 5231

SOUTHWARK DISABLEMENT ASSOCIATION,
Aylesbury Day Centre, 2 Bradenham Close, Lodon SE17 ☎ 0171-701 1391

SERVICECALL SYSTEMS LTD
Millford Lane, Bakewell, Derbyshire DE4 1DX
☎ 01629-812422
Manufacturers of a transmitter/receiver system which enables the user to alert staff at premises where receivers are installed to their need for service. An increasing number of receivers are being installed nationwide in banks, petrol stations, post offices and shops.

SCOPE (formerly the Spastics Society)
12 Park Crescent, London W1N 4EQ. ☎ 0171-636 5020,

SPINAL INJURIES ASSOCIATION
Newpoint House, 76 St James's Lane, London N10 3DF,
☎ 0181-444 2121/883 4296 (counselling line)
National association for people paralysed by spinal cord injury. It provides holiday facilities, including narrow-boats and caravans, adapted for wheelchair users.

WALES COUNCIL FOR THE DISABLED
Llys Ifor, Crescent Road, Caerphilly, Mid Glamorgan CF8 1XL, ☎ 01222-887345
Works closely with the Wales Tourist Board and runs the Access Committee for Wales, an advisory group on all access issues.

YORKSHIRE ASSOCIATION FOR THE DISABLED
St. Georges House, 7-9, Harlow Oval, Harrogate, HG2 0AA,

REVOLUTIONARY CAR SEAT

- It swings around and out
- Entry and exit is made easier
- We supply a kit for many cars to convert the existing seat

FOR DETAILS WRITE TO: **ELAP ENGINEERING LTD**
DEPT OTM1, FORT STREET, ACCRINGTON, LANCS BB5 1QO
TEL 01254 871599 FAX 01254 389992

DON'T BUY....
WITHOUT THE BEST ADVICE.
FREE BOOKLET
EXPLAINS AND GUIDES

POWERCHAIRS
SCOOTERS
RISER CHAIRS
BATHLIFTS
AND MUCH MORE

ESTABLISHED 1881

CALL TODAY
01923 250922

17 Greycaine Rd., Watford, WD2 4JP

TRIPSCOPE

Tripscope, an independent registered charity set up in 1987, provides expert advice, backed up by a computerised information system, on how to get from one place to another for anyone with mobility problems.

The Tripscope staff, all with personal experience of mobility problems, can help plan a journey, whatever the distance and purpose, and however many people are involved. Information is provided on facilities along the way, including wheelchair and scooter hire, and answers will be supplied on any transport- and travel-related question. Tripscope also knows of any special discounts that may be available for disabled travellers.

This information service is free for disabled and elderly people and those who care for them (donations, however, are always welcome in order to maintain and expand the service). To use Tripscope's expertise call one of the numbers given below, or write.

For the UK (except for south-west England and south Wales-and international enquiries, contact: Jim or Adrian, The Courtyard, Evelyn Road, London W4 5JL ☎ 0181-994 9294

For south-west England and south Wales, contact: John, Pamwell House, 160 Pennywell Road, Bristol, Avon BS5 0TX ☎ 0117 941 4024

HOLIDAY SERVICES AND ACCOMMODATION

CALVERT TRUST KIELDER
Kielder Water, Hexham, Northumberland NE48 1BS ☎ 0143-250232 Fax: (01344) 250015
Holiday centre with adventure activities, especially for disabled people.

RITA FISHER
5 Park Lane, Broughton Park, Salford M7 4HT
☎ 0161-792 3029 Fax: 0161 792 3029
Completely organises holidays in Florida, USA for disabled people.

JOHN GROOMS ASSOCIATION FOR DISABLED PEOPLE
10 Gloucester Drive, Finsbury Park, London N42 2LP ☎ 0181-802 7272 Fax: 0181 809 1754
Accommodation and mobile holiday homes for disabled people.

HOLIDAY CARE SERVICE
2 Old Bank Chambers, Station Road, Horley, Surrey RH6 9HW ☎ 01293-774535
Fax: 01293 784647

THE WINGED FELLOWSHIP,
Angel House, 20-32 Pentonville Road, London N1 9XD ☎ 0171-833 2594 Fax: 0171-278 0370
Provide accommodation for disabled people at five centres throughout England.

ORLANDO, FLORIDA

DISABLED TRAVELLERS, DON'T MISS THIS SPACIOUS, LUXURY, ADAPTED BUNGALOW. 3 DOUBLE BEDROOMS, 2 BATHROOMS (MANGAR BATHLIFT), CABLE TV, TELEPHONE, COUNTRY CLUB MEMBERSHIP, 10 MINS AIRPORT, 15 MINS DISNEY, MEET'N'GREET SERVICE AVAILABLE.
PARAPLEGIC OWNER, SUE FISHER,
5 PARK LANE, BROUGHTON PARK,
SALFORD M7 4HT.
TEL/FAX 0161 792 3029

Grooms Holidays

...specialise in integrated holidays for wheelchair users, their families and friends.

- Hotels
- Self-catering units
- Various locations UK and abroad
- Tailor-made holidays
- Luxury coach hire

For further information telephone 0181 800 8695

A division of John Grooms Association for Disabled People
Registered Charity Number 212463

HOMECARE
EQUIPMENT HIRE

HIRE and SALE of:
- Manual wheelchairs
- High seat armchairs and footstools
- Commodes (including chemical & mobile)
- Raised toilet seats & toilet frames
- Bath boards, seats & shower stools
- Back rests, overbed tables & bed cradles
- Bed & chair raisers

DELIVERY SERVICE TO ALL LONDON AREAS
6 days a week
Showroom at: 93 Northcote Road, London, SW11 6 PL
Tel: 0171 924 4058

Useful Products for the Disabled

A J WAY & CO
High Wycombe, Bucks HP12 4HZ
☎ 1494-471821 Fax: 01494-450597
Manufacture motorised seat lifts and reclining chairs for indoor use.

ASHLEY HEALTHCARE LTD
31 Speedwell Road, Hay Mills, Birmingham
B25 8HU ☎ 0121-772 6235

AUTOCHAIR LTD
Milford Lane, Bakewell, Derbyshire DE45 1DX
☎ 01629-813493
Produces the Autochair, which collects and folds the wheelchair from the side of the car and stores it on the roof in a fibreglass enclosure, the Mini-hoist and the Meyland-Carlift, which lifts people from their wheelchair into a car.

BEARD BROTHERS
17 Greycaine Road, Watford, Herts WD2 4JP
☎ 01923-250922 Fax: 01923-238398
Retail outlet for wheelchairs

CARMOBILITY
The Blue Ball Works, Exmouth Road, Colaton Raleigh, near Sidmouth, Devon EX10 0LD
☎ 01395-568830
Manufacturers and distributors of the Carmobility Travel Seat.

COSYFEET
5 The Tanyard, Leigh Road, Street, Somerset
BA16 0HR ☎ 01458-447275 Fax: 01458-45988
Manufacture shoes and slippers for disabled people - particularly for swellings and bunions - which are extremely wide.

CRELLING HARNESS FOR THE DISABLED
11/12 The Crescent East, Cleveleys, Blackpool, Lancashire FY5 3LJ ☎ 01253-852298/821780
Produces belts and harnesses to support and protect disabled children and adults when seated in all models of wheelchairs, pushchairs, and so on. Also designs harnesses for disabled people travelling in vehicles.

ELAP ENGINEERING LTD
Fort Street, Accrington, Lancashire BB5 1QG
☎ 01254-871599
Produces a conversion kit consisting of a swivel mechanism which can be fitted to the basic frame of a car seat enabling the seat to rotate around and beyond the doorway of the vehicle at a right angle.

• ADJ BEDS • COMMODES • ARMCHAIRS • HOISTS •

AT LAST SUPERSTORES FOR THE LESS ABLE

Over 2000 products all under one roof

- Free advice from in-store therapists • Easy Parking & Access
- Open 9-5 pm Daily, Saturday 10-4 pm

HOME ASSESSMENT AVAILABLE ON LARGER ITEMS

FREE Mail Order Catalogue Available

Making life easier

Capital Interchange Way (near Kew Bridge), Brentford, Middlesex TW8 0EX

0181 742 2181

Sterling Park, Pedmore Road, Brierley Hill West Midlands DY5 1TA
(Near the Merry Hill, Shopping Centre)

0384 48 45 44

• CUSHIONS • LIFTERS • POWERCHAIRS • RAMPS • REACHERS •

(Side banners: BACK CARE • SCOOTERS • CUTLERY • WALKING AIDS • SHOWERSEATS • STAIRLIFTS • GRABRAILS • ; TOILET SEATS • BATHLIFTS • COMMODES • TRAYS • RISER CHAIRS • WHEELCHAIRS • KITCHEN AIDS • BATHBOARDS •)

Useful Addresses

HEMCO
Units 58-59 Llandon Industrial Estate,
Cowbridge, South Glamorgan. ☎ 01446-773394
Fax: 01466-772226
Manufacture motorised seat lifts and reclining chairs for indoor use.

HOMECARE EQUIPMENT HIRE
93 Northcote Road, London SW11 6PL
☎.0171-924 4058   Fax: 0171-924 4058
Hire out and sell home equipment for the disabled.

KEEP ABLE
Capital Interchange Way, Brentford, Middlesex
TW8 0EX ☎ 0181 742 2181 Fax: 0181-742 2006
Retail outlet for over 2000 products for the disabled, eg feeding aid, wet weather gear, kitchen utensils.

MOBILITY 2000
Telford Buisiness Centre, Stafford Park 4,
Telford TF3 3BA ☎ 0952-290180
Fax: 0952- 290752

SIMPANTEX HEALTH CARE
Health Care House, 55 Willowfield Road,
Eastbourne, East Sussex BN22 8AP
☎ 01323-411618   Fax: 01323-412781
Manufacture products designed for wheelchair use, such as holdalls.

V & A MARKETING
231A Cathedral Road, Cardiff CF1 9PP
☎ 01222-664564 Fax: 01222-664515
Manufacture an exercise wheelchair called Trim Chair.

WARREN HOOKER
Blackhill Road, Holton Heath Industrial Park,
Poole, Dorset BH16 6LS ☎ 01202-623000
Fax: 01202-623777
Manufacture bathroom products for the disabled.

PROBLEM FEET!

- **BLISSFULLY COMFORTABLE** extra roomy slippers, sandals & shoes for sufferers of swollen feet, bunions, joints, arthritis, etc...

- Probably the **WIDEST, DEEPEST** and **ROOMIEST** footwear available anywhere

- **SHOES** are light as a feather, with soft leather uppers, carefully placed seams and stitching to avoid tender areas, and extra wide deep fitting

- **SLIPPERS** have wide top opening wrap around 'velcro' fastening for adjustable fit, along with extra roomy fitting.

- **COSY FEET** fit feet others won't fit! We supply to hospitals, nursing homes, social services, chiropodists, physiotherapists, etc...

FOR YOUR FREE 20 PAGE COLOUR CATALOGUE

Write to: COSYFEET, (Dept. RAC), 5 The Tanyard, Leigh Road, STREET, Somerset BA16 OHR.
Tel 01458 447275. ACCESS and VISA accepted.

Useful Publications

(See also the section on RADAR.)

British Rail and Disabled Travellers *British Rail* leaflet on facilities for disabled travellers
Care in the Air Advice for disabled travellers *Air Transport Users Committee*
What to do next Disability Rights Handbook *Disability Alliance Educational and Research Association*
Door to Door A guide to transport for people with disabilities *Department of Transport/HMSO*
Flight Plan *Civil Aviation Authority publication for first-time travellers*
Health Advice for Travellers *Department of Health leaflet – Health Literature Line 0800 555 777*
Shopmobility Guidelines *National Federation of Shopmobility*

The Disability Information Trust (see 'Useful addresses') publishes a series of fully illustrated handbooks giving authoritative and independent information on products and ideas to help people with disabilities in their daily lives. Most of the products featured have been tried and assessed by the Trust. For an order form and catalogue call 01865-227592. Titles in the 'Equipment for Disabled People' series include:

Wheelchairs
Outdoor transport
Communication
Clothing and dressing
Housing and home management
Parents with disabilities
Personal care
Gardening

Furniture
Walking aids
Children with disabilities
Hoists and lifts
Arthritis – an equipment guide
Employment and the workplace

THE WINGED FELLOWSHIP,
Angel House,
20-32 Pentonville Road,
London N1 9XD
Tel: 0171 833 2594 Fax: 0171 278 0370

If you're caring and coping for 24 hours a day, seven days a week, a break is not a luxury . . . it's essential.

There are many thousands of families in Britain who have been coping for years without that break.

At our five centres we offer one-to-one caring for severely physically disabled people in a relaxed environment while their carers can choose to enjoy the rest at the centre or elsewhere.

We urgently need voluntary income to support and extend our work as demographic changes increase the demand for respite care.

A legacy or donation could make such a difference.

A Company Limited by Guarantee Reg. No. 2044219. Registered Charity No. 295072.

Places to Visit

The following list of places to visit has been compiled from various sources. The range of disabled facilities available is varied, and we recommend that you telephone in advance the property you wish to visit to confirm that your personal requirements can be met. Where a property is the responsibility of, or sponsored by a particular body, it is indicated by the following abbreviations: National Trust NT; English Heritage EH; Historic Association HHA; Historic Building & Monuments HBM; National Trust for Scotland NTS; Heritage in Wales CADW.

ENGLAND
GREATER LONDON
Cabinet War Rooms, Clive Steps, King Charles St, SW1 2AQ ☎ 0171-930 6961
21 rooms of Winston Churchill's secret underground HQ, kept just as they were during the Second World War, open daily.
Ham House(NT), Ham, Richmond-upon-Thames, TW10 7RS ☎ 0181-940 1950
Hampton Court Palace KT8 9AU
☎ 0181-781 9440
Kew Palace, Kew, TW9 3AB ☎ 0171-781 9540
Osterley Park(NT), Isleworth, TW7 4RB
☎ 0181-560 3918

BEDFORDSHIRE
SILSOE
Wrest Park House & Gardens(EH)
☎ 01525-60152
WOBURN
Woburn Abbey(HHA), MK43 0TP
☎ 01525-290666

BERKSHIRE
READING
Beale Park, Lower Basildon, RG8 9NH
☎ 01734-845172 or 842386
MAIDENHEAD
Cliveden(NT), Taplow, SL6 0JA ☎ 01628-605069
NEWBURY
Highclere Castle(HHA), Highclere Park, RG15 9RN ☎ 01635-253210

BUCKINGHAMSHIRE
BUCKINGHAM
Claydon House(NT), Middle Claydon, MK18 2EY
☎ 01296-730349 or 730693

HIGH WYCOMBE
Hughenden Manor(NT), HP14 4LA
☎ 01494-532580
AYLESBURY
Boarstall Duck Decoy(NT), Boarstall
☎ 01844-237488

CAMBRIDGESHIRE
ELY
Wicken Fen(NT), Lode Ln, Wicken, CB7 5XP
☎ 01353-720274

CHESHIRE
ALTRINCHAM
Dunham Massey Hall(NT), WA14 4SJ
☎ 0161-941 1025
CHESTER
Cheshire Ice Cream Farm, Drumlan Hall, Newton Lane, Tattenhall, CH3 9NE
☎ 01829-70995
Over 30 flavours of ice cream to try, coffee shop offering cream teas, light lunches, etc. Selection

ACCESS FOR ALL WITH THE NATIONAL TRUST

The National Trust, a charity, welcomes visitors with disabilities to its historic buildings, gardens, countryside and coast in England, Wales and Northern Ireland.

1995 is the Trust's Centenary Year and its many events and projects will be accessible to *all* visitors. Details of these, and of access at Trust properties offering enjoyable and comfortable visits is described in a free manual 56-page full colour booklet, sponsored by Fisons plc, and also available in large print. In 1995 a taped version is planned for visually impaired people. Please send a stamped addressed adhesive label (minimum postage) to the address below, stating if large print or tape is needed.

Valerie Wenham, Dept RAC, The National Trust, 36 Queen Anne's Gate, London SW1H 9AS

of rare breeds and farm animals. Free admission, accessible to wheelchairs, ample parking.
CONGLETON
Little Moreton Hall(NT), CW12 4SD
☎ 01260-272018
ELLESMERE PORT
The Boat Museum, Dockyard Road, L65 4EF
☎ 0151-355 5017
An historic dock complex housing over 60 canal craft, with 8 indoor exhibitions, period cottages, blacksmith forge, working engines, boat trips, cafe, shop, parking. Excellent facilities for disabled visitors.
MACCLESFIELD
Capesthorne(HHA), Siddington, SK11 9JY
☎ 01625-861221 or 861779
NORTHWICH
Arley Hall & Gardens(HHA), CW9 6NA
☎ 01565-777353
WILMSLOW
Quarry Bank Mill(NT), Styal, SK9 4LA
☎ 01625-527468

CLEVELAND
MIDDLESBOROUGH
Ormesby Hall(NT), Ormesby, TS7 9AS
☎ 01642-324188

CORNWALL
BODMIN
Lanhydrock(NT), PL30 5AD ☎ 01208-73320
NEWQUAY
Dairyland Farm World, Summercourt, TR8 5AA
☎ 01872-510246
Merry-go-round milking parlour, heritage centre, adventure playground, assault course, nature trail. Lots of farm animals to pet and bottle-feed. Pony rides and tractor rides. Giftshop, teashop and toilets. All areas have disabled access. Open Easter to end October 10.30am-5.30pm.
PENZANCE
Trengwainton Garden(NT), TR20 8RZ
☎ 01736-63021/68410
TRURO
Trewithen House & Gardens, Grampound Rd, TR2 4DD ☎ 01726-882418 or 882585
On the A390 between Truro and St Austell, the outstanding gardens cover about 30 acres and are internationally famous. The nurseries sell a wide range of plants.

CUMBRIA
GRANGE-OVER-SANDS
Holker Hall(HAA), Cark-in-Cartmel, LA11 7PH
☎ 015395-58328
PENRITH
Dalemain(HHA), CA11 0HB ☎ 01768-486450

RAVENGLASS
Muncaster Castle(HHA), CA18 1RQ
☎ 01229-717614
Ravenglass & Eskdale Railway, CA18 1SW
☎ 01229-717171
England's oldest narrow gauge steam railway runs 7 glorious miles from the coast to the foot of the Lakeland's highest hills. Railway museum. Water mill. Bar meals.
ULVERSTON
Laurel & Hardy Museum, 4c Upper Brook Street, LA12 7BH ☎ 01229-582292

DERBYSHIRE
BOLSOVER
Bolsover Castle(EH), Castle St, S42 6PR
☎ 01246-823349
CHESTERFIELD
Hardwick Hall(NT), Doe Lea, S44 5QJ
☎ 02146-850430
MATLOCK
National Tramway Museum, Crich, DE4 5DP
This unique 'Action Stop' offers a mile-long scenic journey through Period Street to open countryside and panoramic views. Tram rides, exhibition hall, video theatre, cafe, shops, playground and picnic areas, plus the largest national collection of vintage electric trams from home and abroad. It's the greatest 'stop' there is for outdoor action with indoor attractions.

DEVON
BARNSTAPLE
Arlington Court(NT), EX31 4LP
☎ 01271-850296
BIDEFORD
Hartland Abbey(HHA), Hartland, EX39 6DT
☎ 01237-41264
COMBE MARTIN
Motorcycle Collection, Cross St, EX39 0DH
☎ 01271-882346
The collection contains British motorcycles, mopeds and power-assisted cycles, displayed against a background of old petrol pumps, signs and garage memorabilia. Motoring nostalgia in an old world atmosphere.
DREWSTEIGNTON
Castle Drogo(NT), EX6 6PB ☎ 01647-433306
PAIGNTON
Paignton & Dartmouth Steam Railway, Queens Park Station, Torbay Road, TQ4 6AF
☎ 01803-555872
Steam trains running for 7 miles in Great Western tradition, along the spectacular Torbay coast to Kingswear for Dartmouth.
PLYMOUTH
Saltram House(NT), Plympton, PL7 3UH
☎ 01752-336546

Places to Visit

TIVERTON
Knightshayes Court & Garden (NT), Bolham, EX16 7RQ ☎ 01884-254665
YELVERTON
Buckland Abbey (NT), PL20 6EY
☎ 01822-853607

DORSET
BEAMINSTER
Parnham House (HHA), DT8 3NA
☎ 01308-862204
DORCHESTER
Athelhampton House (HHA), DT2 7LG
☎ 01305-848363
SWANAGE
Swanage Railway, Station House, BH19 1HB
☎ 01929-425800 or 424276 (talking timetable)
Take a trip back in time, and experience the romance of steam train travel, through the beautiful Purbeck hills, from Swanage to Corfe Castle.
YELVERTON
Paperweight Centre, Leg o' Mutton, PL20 6AD
☎ 01822-854250

CO. DURHAM
DARLINGTON
Railway Centre & Museum, North Rd Station, DL3 6ST ☎ 01325-460532

ESSEX
HARLOW
Mark Hall Cycle Museum, Muskham Road CM20 2LF ☎ 01279-439680
SAFFRON WALDEN
Audley End House & Park (EH), CB11 4JF
☎ 01799-522842

GLOUCESTERSHIRE
BOURTON-ON-THE-WATER
Birdland Park, Riffington Rd, GL54 2BN
☎ 01451-820480
A large collection of exotic birds and colony of penguins. Picnic areas, refreshments, toilet facilities for the disabled. Wheelchair access throughout.
CHELTENHAM
Chedworth Roman Villa (NT), Yanworth, GL54 3LJ ☎ 01242-890256
CHIPPING CAMPDEN
Hidcote Manor Garden (NT), GL55 6LR
☎ 01386-438173
GLOUCESTER
National Waterways Museum, Llanthony Warehouse, Gloucester Docks, GL1 2EH
☎ 01452-318054
200 years of inland waterway history. 'Hands-on' displays, working engines, exhibits, cafe, shop.

Ample car park. Wheelchair friendly. Holders of Gateway Interpret Britain Awards.
TWIGWORTH
Nature in Art, The Centre for International Wildlife Art, Wallsworth Hall, Main A38, GL2 9PA ☎ 01452-731422
The world's first museum dedicated exclusively to art inspired by nature – all media, any country, any period. Parking, ramps, lift, wheelchair users' toilet. Cafe, gardens, shop. Tue-Sun & Bank Hols, 10am-5pm.
WESTBURY-ON-SEVERN
Westbury Court Garden (NT), GL14 1PD
☎ 01452-760461

GREATER MANCHESTER
SALFORD
Lancashire Mining Museum, Buile Hill Park, Eccles Old Road, M6 8GL ☎ 0161-736 1832
Ordsall Hall Museum, M5 3EX
☎ 0161-872 0251
Royal Museum & Art Gallery, Peel Park, The Crescent, M5 4WU ☎ 0161-736 2649
Viewpoint Photography Gallery, The Old Fire Station, The Crescent, M5 4NZ ☎ 0161-737 1040

HAMPSHIRE
ALDERSHOT
Airborne Forces Museum, Browning Barracks, GU11 2BU ☎ 01252-349619
ALRESFORD
Hinton Ampner (NT), Bramdean, SO24 0LA
☎ 01962-771305
ANDOVER
Finkley Down Farm, SP11 6NF ☎ 01264-352195
LYNDHURST
New Forest Museum & Visitor Centre, High St
☎ 01703-283914
PORTSMOUTH
Flagship Portsmouth Trust, The Porter's Lodge, College Rd, HM Naval Base, PO1 3LJ
☎ 01705-861533
ROMSEY
Mottisfont Abbey Garden (NT), SO51 0LP
☎ 01794-341220/340757
SOUTHAMPTON
Lepe Country Park, Lepe, Exbury, SO45 1AD
☎ 01703-899108
Longdown Dairy Farm, Deerleap Ln, Longdown, Ashurst, SO4 4UH
☎ 01703-293326
A must for animal lovers, young and old. Enjoy the hustle and bustle of a busy dairy unit – feed and touch lots of young animals. Refreshments, picnic and play areas, shop.
SOUTHSEA
Pyramids Complex, Clarence Esplanade, PO5 3ST ☎ 01705-877895

Sea Life Centre, Clarence Esplanade
☎ 01705-734461
STOCKBRIDGE
Museum of Army Flying, Middle Wallop, SO20 8DY ☎ 01264-62121
Situated on the A343 adjacent to the operational airfield of the Army Air Corps, the museum houses a unique collection of gliders, aircraft and helicopters, incorporating vivid dioramas depicting the role of Army flying since the early 19th century to the present day. An interesting, educational and fun-packed day for all the family. Free parking, restaurant, cinema, picnic area, open daily 10am-4.30pm. Adult £3.75, OAP £2.75, child £2.25, family £10.
WINCHESTER
Marwell Zoological Park, Colden Common, SO21 1JH ☎ 01962-777406

HEREFORD & WORCESTER
BROMSGROVE
Avoncroft Museum of Buildings, Stoke Heath, B60 4JR ☎ 01527-831886 or 831363
BROMYARD
Lower Brockhampton(NT), WR6 5UH
☎ 01885-488099
DROITWICH
Croft Castle(NT), HR6 9PW ☎ 01568-780246

HERTFORDSHIRE
HATFIELD
Hatfield House, AL9 5NQ ☎ 01707-262823
ROYSTON
Wimpole Hall(NT), Arrington, SG8 0BW
☎ 01223-207257
WELWYN
Shaw's Corner(NT), Ayot St Lawrence, AL6 9BX
☎ 01438-820307

HUMBERSIDE
DRIFFIELD
Sledmere House, Sledmere, YO25 0XG
☎ 01377-236208

ISLE OF WIGHT
NEWPORT
Carisbrooke Castle, PO30 1XY ☎ 01983-522107
WROXALL
Appuldurcombe House(EH), PO38 3EW
☎ 1983-852484

KENT
CRANBROOK
Sissinghurst Castle Gardens(NT), Sissinghurst, TN17 2AB ☎ 01580-712850
EDENBRIDGE
Hever Castle, Hever, TN8 7NG ☎ 01732-865224

HYTHE
Port Lympne Zoo Park, Mansion and Gardens, Port Lympne, CT21 4PD ☎ 01303-264646
MAIDSTONE
Leeds Castle, ME17 1PL ☎ 01622-765400
PADDOCK WOOD
Whitbread Hop Farm, Beltring, TN12 6PY
☎ 01622-872068
SEVENOAKS
Ightham Mote(NT), Ivy Hatch, TN15 0NT
☎ 01732-810378

LANCASHIRE
BOLTON
Turton Tower, Chapeltown Rd, Turton, BL7 0HG
☎ 01204-852203
BURNLEY
Gawthorpe Hall(NT), Padiham, BB12 8UA
☎ 01282-78511
CLITHEROE
Brownsholme Hall, BB7 3DE ☎ 01254-826719

LEICESTERSHIRE
MARKET BOSWORTH
Bosworth Battlefield Visitor Centre, CV13 0AD
☎ 01455-290429
OAKHAM
Oakham Castle, Market Place, LE15 6HW
☎ 01572-723654

LINCOLNSHIRE
WESTON
Baytree Owl Centre, High Road, PE12 6JU ☎ 01406-371907
A large, varied collection of owls from around the world including Gracie, the centre's star, and Kelly, the secretary bird. Daily flying displays, glasshouse, gift shop. £1.80, OAP/child £1.30.

MERSEYSIDE
LIVERPOOL
Speke Hall(NT), The Walk, L24 1XD
☎ 0151-427 7231

NORFOLK
CROMER
Lifeboat Museum and Lifeboat Station, The Gangway. ☎ 01263 512503
History of the Lifeboat Station from 1804 and includes the lifeboat HS Bailey which served on the station from 1935 to 1945 and saved over 800 lives. Also models and the medals awarded to Henry Blogg. School parties welcome; educational room with TV and video, and lecture room. Easter-end Oct, 10-4pm daily.
KING'S LYNN
Houghton Hall, Houghton, PE31 6UE
☎ 01485-528569

Lynn Museum, Old Market Street, PE30 1NL
☎ **01553-775001**
Oxburgh Hall(NT), Oxborough, PE33 9PS
☎ 01366-328258

NORWICH
Blickling Hall(NT), Aylsham, NR11 6NF
☎ 01263-733084
Norwich Castle, NR1 3JU ☎ 01603-223624

SHERINGHAM
Sheringham Park(NT), Upper Sheringham,
NR26 8TB ☎ 01263-823778

NORTHAMPTONSHIRE

DAVENTRY
Canons Ashby House(NT), NN11 6SD
☎ 01327-860044

KETTERING
Boughton House(HHA), NN14 1BJ
☎ 01536-515731

NORTHUMBERLAND

MORPETH
Cragside House(NT), Rothbury, NE65 7PX
☎ 01669-20333 or 20266

ROTHBURY
Bamburgh Castle, NE65 7SP ☎ 01669-20314

STOCKFIELDS
Cherryburn(NT), Station Bank, Mickley, NE43 7DB ☎ 01661-843276

NOTTINGHAMSHIRE

NEWARK
Carlton Hall(HAA), Carlton-on-Trent, NG23 6NW ☎ 01636-821421

NOTTINGHAM
Holme Pierrepont Hall(HHA), Radcliffe-on-Trent, NG12 2LD ☎ 01607-332371

WORKSOP
Clumber Park(NT), S80 3AZ ☎ 01909-476592

OXFORDSHIRE

BANBURY
Upton House(NT), OX15 6HT ☎ 01295-670266

WOODSTOCK
Blenheim Palace(HHA), OX20 1PX
☎ 01993-811091

SHROPSHIRE

BRIDGNORTH
Dudmaston(NT), Quatt, WV15 6QN
☎ 01746-780866

OSWESTRY
Tyn-y-Rhos Hall, Weston Rhyn, SY10 7NQ
☎ 01691-777898

SHIFNAL
Weston Park, Weston-under-Lizard, TF11 8LE
☎ 01952-76207 or 76385

SHREWSBURY
Attingham Park(NT), SY4 4TP ☎ 01743-709203

SOMERSET

MINEHEAD
Dunster Castle(NT), Dunster, TA24 6SL
☎ 01643-821314

STAFFORDSHIRE

STAFFORD
Shugborough(NT), Milford, ST17 0XB
☎ 01889-881388

WOLVERHAMPTON
Moseley Old Hall(NT), Moseley Old Hall Ln,
Fordhouses, WV10 7HY ☎ 01902-782808

SUFFOLK

BURY ST EDMUNDS
Ickworth(NT), The Rotunda, Horringer, IP29 5QE ☎ 01284-735270

LOWESTOFT
Somerleyton Hall, NR32 5QQ ☎ 01502-730224

NEWMARKET
National Horseracing Museum, 99 High Street, CB8 8JL ☎ 01638-667338
400 years of horseriding history in paintings, sculptures, trophies, memorabilia and graphic displays. Video and computer information. Award-winning licensed coffee shop, and excellent gift shop.

SUDBURY
Melford Hall(NT), Long Melford, CO10 9AH
☎ 01787-880286

SURREY

DORKING
Polesden Lacey(NT), RH5 6BD ☎ 01372-458203 or 452048

FARNHAM
Rural Life Centre, Reeds Rd, Tilford, GU10 2DL
☎ 01252-795571/792300

GUILDFORD
Clandon Park(NT), West Clandon, GU4 7RQ
☎ 01483-222482
Hatchlands Park(NT), East Clandon, GU4 7RT
☎ 01483-222482

SUSSEX

ASHINGTON
Holly Gate Cactus Nursery & Gardens, Billingshurst Road, RH20 3BA ☎ 01903-892930

BATTLE
Battle Abbey(EH), High St, TN33 0AD
☎ 01424-773792

BEXHILL-ON SEA
Bexhill Museum of Costume and Social History, The Manor Gardens, Upper Sea Road, TN40 1RL ☎ 01424-210045

Costumes from 1740, dolls, toys, household effects. Free car park. Adults £1, children 50p.
CHICHESTER
Goodwood House (HHA), Goodwood, PO18 0PX ☎ 01243-774107
HAYWARDS HEATH
Nymans Garden (NT), Handcross, RH17 6EB ☎ 01444-400002
LEWES
Bentley Wildfowl & Motor Museum, Halland BN8 5AF ☎ 01825-840573
A wildlife reserve, a stately home and a motor museum. Also an adventure playground, picnic area, shop and licensed tearoom. Steam railway and free parking.
PETWORTH
Petworth House (NT), GU28 0AE ☎ 01798-342207
WORTHING
Worthing Museum & Art Gallery, Chapel Road, BN11 1HP ☎ 01903-239999

WARWICKSHIRE
ALCESTER
Coughton Court (NT), B49 5JA ☎ 01789-762435
SOLIHULL
Baddesley Clinton (NT), Knowle, B93 0DQ ☎ 01564-783294
Packwood House (NT), Lapworth, B94 6AT ☎ 01564-782024
STRATFORD-UPON-AVON
Shakespeare Birthplace Trust, Henley St, CV37 6QW ☎ 01789-204016
WARWICK
Charlecote Park (NT), Wellsbourne, CV35 9ER ☎ 01789-470277

WILTSHIRE
AVEBURY
Avebury Manor (NT), SN8 1RF ☎ 01672-539250
CHIPPENHAM
Dyrham Park (NT), Dyrham, SN14 8ER ☎ 01272-372501
Lacock Abbey (NT), Lacock, SN15 2LG ☎ 01249-730459
SALISBURY
Wilton House, SP2 0BJ ☎ 01722-734115
WARMINSTER
Longleat House, BA12 7NW ☎ 01985-844400

WEST MIDLANDS
DUDLEY
Dudley Zoo and Castle, Castle Hill, DY1 4RB ☎ 01384-252401

YORKSHIRE
BRADFORD
Cartwright Hall Art Gallery, Lister Park, BD9 4NS ☎ 01274-493313
HALIFAX
Shibden Hall & Folk Museum, HX3 6XG ☎ 01422-352246
HELMSLEY
Rievaulx Terrace & Temples (NT), Rievaulx, YO6 5LJ ☎ 0143-96340
KEIGHLEY
East Riddlesden Hall (NT), Bradford Rd, BD20 5EL ☎ 01535-607075
PICKERING
Pickering Castle (EH), YO18 7AX ☎ 01751-74989
RIPON
Newby Hall & Gardens, Newby, HG4 5AE ☎ 01423-322583
Renowned Adam house, unique tapestries, classical statuary, award-winning gardens, children's adventure garden, miniature railway, restaurant, shop, plant stall. Five wheelchairs available, good disabled facilities and mapped route through gardens.
SCARBOROUGH
Scarborough Castle (EH), Castle Rd, YO1 1HY ☎ 01723-372451
THIRSK
Sion Hill Hall (HHA), Kirby Wiske ☎ 01845-587206
WAKEFIELD
Nostell Priory (NT), Doncaster Rd, Nostell, WF4 1QE ☎ 01924-863892
YORK
Beningbrough Hall (NT), Shipton-by-Beningbrough, YO6 1DD ☎ 01904-470666
Castle Howard, YO6 7BZ ☎ 01653-684444

SCOTLAND
BORDERS

AYTON
Ayton Castle (HHA), TD14 5RD ☎ 018907-81212
MELROSE
Abbotsford House ☎ 01896-752043
SELKIRK
Bowhill, TD7 5ET ☎ 01750-20732

FIFE
CUPAR
Hill of Tarvit mansion house (NTS) ☎ 01334-53127
PITTENWEEM
Kellie Castle (NTS) ☎ 0133-38271

GRAMPIAN
ABERDEEN
Drum Castle (NTS) ☎ 01330-811204
BANCHORY
Crathes Castle (NTS) ☎ 01330-44525
KENNETHMONT
Leith Hall & Garden (NTS) ☎ 01464-831216

METHLICK
Haddo House (NTS) ☎ 01651-851440
NAIRN
Brodie Castle (NTS) ☎ 01309-641371

HIGHLAND
ARDESIER
Fort George (HBM) ☎ 01667-2777
GOLSPIE
Dunrobin Castle & Gardens, KW10 6RR
☎ 01408-633177 or 633268
ISLE OF SKYE
Dunvegan Castle, IV55 8WF ☎ 01470-22206
KYLE OF LOCHALSH
Eilean Donan Castle, Dornie ☎ 01599-555202

LOTHIAN
EDINBURGH
Edinburgh Castle (HBM), Castle Hill, EH1 ☎ 0131-225 9846
LINLITHGOW
The House of Binns (NTS) ☎ 01506-834255
SOUTH QUEENSFERRY
Dalmeny House, EH30 9TQ ☎ 0131-331 1888

STRATHCLYDE
MAYBOLE
Culzean Castle, Garden & Country Park (NTS), KA19 8LE ☎ 0165-56274
AYR
Burns Cottage, Alloway, KA7 4PY
☎ 01292-441215
ISLE OF ARRAN
Brodick Castle, Garden & Country Park (NTS)
☎ 01770-302202
FAIRLIE
Kelburn Castle, KA29 0BE ☎ 01475-568685

TAYSIDE
EDZELL
Edzell Castle & Gardens (HBM), ☎ 0135-648631
PERTH
Scone Palace, PH2 6BD ☎ 01738-52300

WALES
CLWYD
BODELWYDDAN
Bodelwyddan Castle, St Asaph LL18 5YA
☎ 01745-584060

GWENT
TINTERN
Tintern Abbey (CADW), ☎ 01291-689251

GWYNEDD
BANGOR
Penrhyn Castle (NT), LL57 4HN
☎ 01248-353084

POWYS
WELSHPOOL
Powis Castle (NT), SY21 8RF ☎ 01938-554336

NORTHERN IRELAND
CO ANTRIM
TEMPLEPATRICK
Patterson's Spade Mill (NT), Antrim Rd, BT29 0AO ☎ 01849-433619

CO ARMAGH
PORTADOWN
Ardress House (NT), 64 Ardress Rd, BT62 1SQ
☎ 01762-851236

CO DOWN
BALLYNAHINCH
Rowallane (NT), Saintfield, BT24 7LH
☎ **01238-511242**
DOWNPATRICK
Castle Ward (NT) Strangford, BT30 7LS
☎ 01396-881204
NEWTOWNARDS
Mount Stewart (NT), BT22 2AD
☎ 012477-88387 or 88487

CO FERMANAGH
ENNISKILLEN
Castle Coole (NT) ☎ 01365-322690
Florence Court (NT) ☎ 01365-82249
NEWTOWNBUTLER
Crom Estate (NT), BT74 ☎ 013657-38174

CO LONDONDERRY
CASTLEROCK
Hezlett House (NT), 107 Sea Rd, Liffock, BT51 4RE ☎ 01265-848567
MAGHERAFELT
Springhill (NT), 20 Springhill Rd, Moneymore, BT45 7NQ ☎ 01648-748210

CO TYRONE
DUNGANNON
The Argory (NT), Moy, BT71 6NA
☎ 01868-784753

REPUBLIC OF IRELAND
CO CLARE
CRATLOE
Cratloe Woods House ☎ 061-327028
QUIN
Knappogue Castle ☎ 061-71103

CO CORK
BLARNEY
Blarney Castle ☎ 021-385252

PLACES TO VISIT

CARRIGTWOTHILL
Fota House ☎ 021-812555

CO DUBLIN

DUBLIN
Powerscourt Townhouse Centre, 59 South William St ☎ 01-687477

SANDYCOVE
James Joyce Tower ☎ 02-809265 or 808571

CO GALWAY

CONNEMARA
Kylemore Abbey, Kylemore ☎ 095-41146

GORT
Thoor Ballylee ☎ 091-31436

CO KERRY

KILLARNEY
Muckross House & Gardens ☎ 064-31440

CO KILDARE

CELBRIDGE
Castletown House ☎ 01-288252

CO LAOIS

PORTLAOISE
Emo Court ☎ 0502-26110

CO LIMERICK

GLIN
Glin Castle ☎ 068-34173 or 34112

RATHKEALE
Castle Matrix ☎ 069-64284

CO LONGFORD

LONGFORD
Carrigglas Manor ☎ 043-45165

CO ROSCOMMON

CASTLEREA
Clonalis House ☎ 0907-20014

CO SLIGO

SLIGO
Lissadell ☎ 071-63150

CO WEXFORD

WEXFORD
The Irish Agricultural Museum ☎ 053-42888

CO WICKLOW

BLESSINGTON
Russborough ☎ 045-65239

BRAY
Killruddery House & Gardens ☎ 1-863405

ONE OF THE WORLD'S LEADING AVIATION MUSEUMS HAS A NEW ATTRACTION THAT IS DIFFERENT

CATCH THE SPIRIT OF NAVAL AVIATION ON THE ULTIMATE CARRIER EXPERIENCE

Yes, a flight deck on land! all the sights, smells, sounds, actions and activity of a ship from the Big Carrier era - on a mission of mercy. This unique experience is open and waiting for you to come aboard.
Summer: Last flights to Carrier depart at 4pm. Museum closes at 5.30pm. *Winter:* Last flights to Carrier depart at 3pm. Museum closes at 4.30pm.

Other exhibits:
- Concorde 002
- World Wars 1 & 2
- Kamikaze ■ Wrens
- Korean War
- Recent Conflicts
- Harrier 'Jump' Jet

Facilities:
- Licensed restaurant
- Children's adventure playground
- Gift shop ■ Free parking
- Airfield viewing galleries
- Baby care room
- Access for the disabled to all areas

FLEET AIR ARM MUSEUM

RACOTM

RNAS Yeovilton, Ilchester, Somerset, BA22 8HT Tel: (01935) 840565 Fax: (01935) 840181

Places to Stay

The following pages include the RAC Appointed and Listed hotels which, in the managers opinion, can cater for disabled people. The facilities available differ from hotel to hotel, and it is important to ask the hotel whether they can fulfil your requirements.

Where a hotel has informed us of the facilities they can offer, these are shown. The full list is: designated parking space, ramp or level access, doors manageable by someone in a wheel chair, lift/specifically adapted lift for wheelchairs, specially adapted ground-floor bedrooms, adapted en suite bathrooms/adapted bathrooms.

In addition, the list includes camping and caravanning sites in the RAC guide which offer facilities for the disabled.

The list is arranged by country, then alphabetically by town.

SYMBOLS AND ABBREVIATIONS USED

      ♿   facilities for the disabled
      ♘   country house hotel
    sB&B   price for bed and breakfast for one person for one night
     sB   price for room only for one person for one night

RAC HOTEL AWARDS

          Blue Ribbon for all round excellence
      H   Hospitality award for service
      C   Comfort award
      R   Award for Restaurant excellence
          RAC Farm
          RAC Campsite

England

ABINGDON *Oxfordshire*
★★★**Abingdon Lodge Hotel** Marchám Road, Abingdon OX14 1TZ 01235-553456 Fax 01235-554117
sB&B £77.95

★★★**Upper Reaches Hotel** Thames Street, Abingdon OX14 3JA 01235-522311 Fax 01235-555182
sB&B £88.50

ADSTOCK *Buckinghamshire*
(Inn) **The Folly** *Highly Acclaimed* Buckingham Road, Adstock MK18 2HS 01296-712671
sB&B £24

ALDEBURGH *Suffolk*
★★★**Brudenell Hotel** The Parade, Aldeburgh IP15 5BU 01728-452071 Fax 01728-454082
sB&B £73-£80

ALDERLEY EDGE *Cheshire*
★★★H **Alderley Edge** Macclesfield Road, Alderley Edge SK9 7BJ 01625-583033 Fax 01625-586343
sB&B £95.50

ALFRISTON *East Sussex*
★★★**Star Inn** High Street, Alfriston BN26 5TA 01323-870495 Fax 01323-870922
sB&B £78.50

ALSTON *Cumbria*
★★**Nent Hall Hotel** Alston CA9 3LQ 01434-381584 Fax 01434-382668
sB&B £50

ALTRINCHAM *Cheshire*
★★★H **Cresta Court Hotel** Church Street, Altrincham WA14 4DP 0161-927-7272 Fax 0161-926-9194
sB&B £55.50

ALVESTON *Avon*
★★★HCR **Alveston House Hotel** Alveston BS12 2LJ 01454-415050 Fax 01454-415425
sB&B £67.50-£71.50

AMBLESIDE *Cumbria*
★★★HC **Rothay Manor Hotel** Rothay Bridge, Ambleside LA22 0EH 0153-94-33605 Fax 0153-94-33607
sB&B £72

★★CR **Borrans Park Hotel** Borrans Road, Ambleside LA22 0EN 0153-94-33454
sB&B £29-£39

♿ ★★HC **Kirkstone Foot Hotel** Kirkstone Pass Road, Ambleside LA22 9EH 0153-94-32232 Fax 0153-94-32232
sB&B £49.50-£62

Rowanfield Country House *Highly Acclaimed* Kirkstone Road, Ambleside LA22 9ET 0153-94-33686
sB&B £48-£52

▲☐ **Skelwith Fold Caravan Park** Ambleside

LA22 0HX 015394-32277
Car & Caravan £8

AMPFIELD *Hampshire*
★★★**Potters Heron Hotel** Ampfield SO51 9ZF 01703-266611 Fax 01703-251359
sB&B £72.50

APPLEBY-IN-WESTMORLAND *Cumbria*
ℚ ★★★**Appleby Manor Hotel** Roman Road, Appleby-in-Westmorland CA16 6JB 0176-83-51571 Fax 0176-83-52888
sB&B £58-£64

ARUNDEL *West Sussex*
★★★**Arundel Resort** 16 Chichester Road, Arundel BN18 0AD 01903-882677
Fax 01903-884154
sB&B £62.50

★★★**(Inn) Howards Hotel** Crossbush, Arundel BN19 9PQ 01903-882655 Fax 01903-883384
sB&B £47.50-£49.50

Arundel Park Inn & Travel Lodge *Acclaimed* The Causeway, Station Approach, Arundel BN18 9JL 01903-882588 Fax 01903-883808
sB&B £32-£38

▲⚐ **Maynard's Caravan & Camping Park**
Crossbush, Arundel BN18 9PQ 01903-882075
Car & Caravan £8.50

ASHBOURNE *Derbyshire*
▲⚐ **Gateway Caravan Park** Osmaston, Ashbourne 01335-44643/43
Car & Caravan £8

ASHBURTON *Devon*
ℚ ★★★**Holne Chase Hotel** Ashburton TQ13 7NS 01364-631471 Fax 01364-631453
sB&B £47.50

★★**Dartmoor Lodge** Pear Tree Cross, Ashburton TQ13 7JW 01364-652232 Fax 01364-653990

Two storey country hotel.
♿ parking, ramp (or level) access, wheel chair Lift, adapted en-suite bathrooms, sB&B £38.50

🚜 **Higher Mead Farm** Ashburton TQ13 7LJ 01364-652598

▲⚐ **River Dart Country Park** Holne Park, Ashburton TQ13 7NP 01364-652511
Car & Caravan £4

ASHBY-DE-LA-ZOUCH *Leicestershire*
★★★**The Fallen Knight** Kilwardy Street, Ashby-de-la-Zouch 01530-412230 Fax 01530-417596
sB&B £62

ASHFORD *Kent*
★★★★**Ashford International Hotel** Simone Weil Avenue, Ashford TN24 8UX 01233-611444
Fax 01233-627708

A modern hotel with an attractive glass covered boulevard with adjoining shops and restaurants.
♿ramp (or level) access, wheel chair Lift, adapted en-suite bathrooms, sB&B £89.75

ASHINGTON *Northumberland*
▲⚐ **Wansbeck Riverside Park** Ashington 01233-814444
Car & Caravan £6.80

ASHTON-UNDER-LYNE *Gtr Manchester*
Welbeck House Hotel *Acclaimed* 324 Katharine Street, Ashton-under-Lyne OL6 7BD
0161-3440751
sB&B £29.50

ASTON CLINTON *Buckinghamshire*
★★★*HCR* **Bell Inn** London Road, Aston Clinton HP22 5HP 01296-630252 Fax 01296-631250
17th century coaching house, with a rose covered entrance and many fine original features. Originally a coaching stop for Duke of Buckingham.
♿ramp (or level) access, sB&B £79

AXMINSTER *Devon*
▲⚐ **Andrewshayes Caravan Park** Dalwood, Axminster EX13 7DY 01297-831225
Car & Caravan £8

AYLESBURY *Buckinghamshire*
🚜 ℚ ★★★★**Hartwell House Hotel**

PLACES TO STAY

Oxford Road, Aylesbury HP17 8NL 01296-747444
Fax 01296 747450
sB&B £107.50

BAMPTON *Oxon*
The Farmhouse *Highly Acclaimed* University Farm, Lew, Bampton OX18 2AU 01993-850297
Fax 01993-850965
sB&B £39

BANHAM *Norfolk*
Farm Meadow Caravan & Camping Park The Grove, Banham Zoo, Banham NR16 2HB 01953-887771
Car & Caravan £6

BARNSTAPLE *Devon*
★★★**Barnstaple Hotel** Braunton Road, Barnstaple EX31 1LE 01271-76221
Fax 01271-24101
sB&B £47

★★★**Royal & Fortescue Hotel** Boutport Street, Barnstaple EX31 3HG 01271-42289
Fax 01271-42289
sB&B £47

Cedars Lodge Inn Bickington Road, Barnstaple EX31 2HP 01271-71784
sB&B £37

BARROW IN FURNESS *Cumbria*
★★★★**Abbey House Hotel** Abbey Road, Barrow In Furness LA13 0PA 01229-838282
Fax 01229-820403
The Abbey House takes its name from the neighbouring ruins of Furness Abbey. It is a fine, sandstone mansion designed by Sir Edward Lutyens and is surrounded by lawns and wooded areas.
ramp (or level) access, adapted en-suite bathrooms, sB&B £77.20

BASILDON *Essex*
Campanile Miles Gray Road, Pipps Hill, Basildon SS14 3AE 01268-530810 Fax 01268-286710
sB&B £40

BASINGSTOKE *Hampshire*
★★★★*CR* **Audleys Wood Thistle Hotel** Alton Road, Basingstoke RG25 2JT 01256-817555
Fax 01256-817500
sB&B £92.75

BATH *Avon*
★★★★★**Bath Spa Hotel** Sydney Road, Bath BA2 6JF 01225-444424 Fax 01225-444006
sB&B £121.75-£141.75

★★★★**Royal Crescent Hotel** Royal Crescent, Bath BA1 2LS 01225-319090
Fax 01225-339401
sB&B £108.25

BATTLE *East Sussex*
★★★ *R* **Powdermills Hotel** Powdermills Lane, Battle TN33 0SP 01424-775511 Fax 01424-774540
sB&B £45

Netherfield Hall Netherfield, Battle TN33 9PQ 01424-774450
sB&B £20-£25

BEACONSFIELD *Buckinghamshire*
★★★★**Bell House Hotel** Oxford Road, Beaconsfield HP9 2XE 01753-887211
Fax 01753-888231
sB&B £110

BEANACRE *Wiltshire*
★★★*HCR* **Beechfield House Hotel** Beanacre SN12 7PU 01225-703700 Fax 01225-790188
sB&B £55

BEAULIEU *Hampshire*
★★★**Master Builders House** Bucklers Hard, Beaulieu SO42 7XB 01590-616253
Fax 01590-616297
sB&B £55

BEDALE *North Yorkshire*
Elmfield Country House *Highly Acclaimed* Arrathorne, Bedale DL8 1NE 01677-450558
Fax 01677-450557
sB&B £27-£29.50

BEDFORD *Bedfordshire*
★★★**(Inn) The Wayfarer** 403 Goldington Road, Bedford MK41 0DS 01234-272707
Fax 01234-272707
sB&B £55

★★**(Inn)** *R* **Knife & Cleaver Hotel** The Grove, Houghton Conquest, Bedford MK45 3LA 01234-740387 Fax 01234-740900
sB&B £41-£51

BELFORD *Northumberland*
★★★*C* **Blue Bell Hotel** Market Place, Belford NE70 7ND 01668-213543 Fax 01668-213787
sB&B £42

★★★*HC* **Waren House Hotel** Waren Mill, Belford NE70 7EE 01668-214581
Fax 01668-214484
sB&B £74

BELLINGHAM *Northumberland*
★★*HR* **Riverdale Hall Hotel** Hexham, Bellingham NE48 2JT 01434-220254
Fax 01434-220457
sB&B £40-£45

BELPER *Derbyshire*
Shottle Hall Shottle, Belper DE5 2EB 01773-550203
sB&B £23-£27

BERKSWELL *Warwickshire*
★★★*CR* **Nailcote Hall Hotel** Nailcote Lane, Berkswell CV7 7DE 01203-466174 Fax 01203-470720
sB&B £95

BEXLEYHEATH *Kent*
★★★★**Swallow Hotel** 1 Broadway, Bexleyheath DA6 7JZ 0181-298-1000 Fax 0181-298-1234
sB&B £88

BIDEFORD *Devon*
♉ ★★★*HC* **Portledge Hotel** Fairy Cross, Bideford EX39 5BX 01237-451262
Fax 01237-451717
sB&B £32-£42

★★*H* **Riversford Hotel** Limers Lane, Bideford EX39 2RG 01237-474239 Fax 01237-421661
sB&B £37

Pines *Acclaimed* Eastleigh, Bideford EX39 4PA 01271-860561 Fax 01271-860561
sB&B £20-£25

BIRCHINGTON *Kent*
⛺🚐 **Two Chimneys Caravan Park** Five Acres, Shottendane Road, Birchington CT7 0HD 01843-41068
Car & Caravan £12

BIRKENHEAD *Merseyside*
★★★**Bowler Hat Hotel** Talbot Road, Oxton, Birkenhead L43 2HH 0151-6524931
Fax 0151-653-8127
sB&B £59.50

♉ ★★*R* **Riverhill Hotel** Talbot Road, Oxton, Birkenhead L43 2HJ 0151-653-3773
Fax 0151-653-7162
sB&B £44

BIRMINGHAM *West Midlands*
★★★★★**Swallow Hotel** 13 Hagley Road, Birmingham B16 8SJ 0121-452-1144
Fax 0121-456-3442
sB&B £120

★★★★**Copthorne Hotel** Paradise Circus, Birmingham B3 3HJ 0121-200-2727
Fax 0121-200-1197
sB&B £112.25

★★★★**Holiday Inn** Central Square, Birmingham B1 1HH 0121-631-2000 Fax 0121-643-9018
sB&B £90-£109

★★★**Grand Hotel** Colmore Row, Birmingham B3 2DA 0121-236-7951 Fax 0121-233-1465
sB&B £89.50

★★★**Novotel** 70 Broad Street, Birmingham B1 2MT 0121-643-2000 Fax 0121-643-9796
sB&B £74.50

★★★*C* **Westley Arms Hotel** Westley Road, Acocks Green, Birmingham B27 7UF 0121-706-4312 Fax 0121-706-2824
sB&B £64

Chamberlain Hotel *Highly Acclaimed* Alcester Road, Birmingham B12 0TJ 0121-6270627
Fax 0121-6270628
sB&B £35

Lyndhurst Hotel *Acclaimed* 135 Kingsbury Road, Erdington, Birmingham B24 8QT 0121-3735695
Fax 0121-3735695
sB&B £36.50-£39.50

Campanile 55 Irving Street, Birmingham B1 1DH 0121-622-4925 Fax 0121-622-4195
sB&B £40

BISHOPS TAWTON *Devon*
♉ ★★**Downrew House Hotel** Barnstaple, Bishops Tawton EX32 0DY 01271-42497
Fax 01271-23947
sB&B £38.50-£45.50

BLACKBURN *Lancashire*
★★★**Blackburn Moat House** Yew Tree Drive, Blackburn BB2 7BE 01254-264441
Fax 01254-682435
sB&B £57.50

★★★**(Inn)** *C* **Mytton Fold Farm** Whalley Road, Langho, Blackburn BB6 8AB 01254-240662
Fax 01254-248119
sB&B £45

BLACKPOOL *Lancashire*
★★★★**Imperial Hotel** North Promenade, Blackpool FY1 2HB 01253-23971
Fax 01253-751784
sB&B £55-£69

★★★★*R* **Pembroke Hotel** North Promenade, Blackpool FY1 2JQ 01253-23434 Fax 01253-27864
sB&B £105

★★★**Clifton Hotel** Talbot Square, Promenade, Blackpool FY1 1ND 01253-21481
Fax 01253-27345
sB&B £25-£75

★★**Claremont Hotel** 270 North Promenade, Blackpool FY1 1SA 01253-293122
Fax 01253-752409
sB&B £30-£70

★★**Kimberley Hotel** New South Promenade, Blackpool FY4 1NQ 01253-341184
Fax 01253-408737
sB&B £26.35

Brooklands Hotel *Highly Acclaimed* 28-30 King Edward Avenue, North Shore, Blackpool FY2 9TA 01253-51479
sB&B £16-£18

Burlees Hotel *Highly Acclaimed* 40 Knowle Avenue, North Shore, Blackpool FY2 9TQ 01253-354535
sB&B £20-£24

Mimosa Hotel *Acclaimed* 24a Lonsdale Road, Blackpool FY1 6EE 01253-341906
sB&B £16

⚠🚐 **Kneps Farm Holiday Home & Park** River Road, Thornton Cleveleys, Blackpool FY5 5LR 01253-823632

⚠🚐 **Marton Mere Caravan Park** Mythop Road, Blackpool FY4 4XN 01253-767544
Car & Caravan £9.95

⚠🚐 **Stanah House Caravan Park** River Road, Thornton, Blackpool FY5 5LR 01253-824000
Car & Caravan £7

BLAKENEY *Norfolk*
★★★**Blakeney Hotel** Quayside, Blakeney NR25 7NE 01263-740797 Fax 01263-740795
sB&B £46-£55

★★**Manor Hotel** Holt, Blakeney NR25 7ND 01263-740376 Fax 01263-741116
sB&B £30

BLANDFORD FORUM *Dorset*
⚠🚐 **The Inside Park Caravan & Camping** Blandford Forum DT11 0HG 01258-453719
Fax 01258-454026

BOLDON *Tyne & Wear*
★★★**Friendly Hotel** Witney Way, Jct A19/A184, Boldon NE35 9PE 0191519-1999
Fax 0191519-0655
sB&B £67.90-£81.10

BOLNEY *West Sussex*
★★★**Hickstead Resort Hotel** Jobs Lane, Bolney RH17 5PA 01444-248023 Fax 01444-245380
sB&B £72.50

BOLTON *Gtr Manchester*
★★★*HC* **Bolton Moat House** 1 Higher Bridge Street, Bolton BL1 2EW 01204-383338
Fax 01204-380777
sB&B £90

BOLTON ABBEY *North Yorkshire*
★★★*HCR* **Devonshire Arms Country House** Bolton Abbey BD23 6AJ 01756-710441
Fax 01756-710564
sB&B £90-£100

BOROUGHBRIDGE *North Yorkshire*
★★★**Crown Hotel** Horsefair, Boroughbridge YO5 9LB 01423-322328 Fax 01423-324512
sB&B £45

BOSCASTLE *Cornwall*
Old Coach House Tintagel Road, Boscastle PL35 0AS 01840-250398
Beautiful, 300 year old, former coaching house.
♿ parking, ramp (or level) access, adapted en-suite bathrooms, sB&B £15-£24

BOURNEMOUTH *Dorset*
★★★★*HCR* **Norfolk Royale Hotel** Richmond Hill, Bournemouth BH2 6EN 01202-551521
Fax 01202-299729
sB&B £98.75

★★★**Anglo Swiss Hotel** 16 Gervis Road, East Cliff, Bournemouth BH1 3EQ 01202-554794
Fax 01202-299615
sB&B £40

★★★**Belvedere Hotel** Bath Road, Bournemouth BH1 2EU 01202-297556 Fax 01202-294699
sB&B £22-£45.50

★★★**Bournemouth Heathlands** 12 Grove Road, East Cliff, Bournemouth BH1 3AY 01202-553336
Fax 01202-555937
sB&B £49-£54

★★★*H* **Burley Court Hotel** Bath Road, Bournemouth BH1 2NP 01202-552824
Fax 01202-298514
sB&B £29-£42

★★★**Cliffside Hotel** East Overcliff Drive, Bournemouth BH1 3AQ 01202-555724
Fax 01202-555724
sB&B £37.50-£44.50

★★★*C* **Connaught Hotel** West Hill Road, Bournemouth BH2 5PH 01202-298020
Fax 01202-298028
sB&B £39-£41

★★★**Courtlands Hotel** 16 Boscombe Spa Road, Bournemouth BH5 1BB 01202-302442
Fax 01202-309880
sB&B £44

★★★**Durley Hall Hotel** Durley Chine Road, Bournemouth BH2 5JS 01202-500100
Fax 01202-500103
sB&B £32-£43

★★★**Durlston Court Hotel** Gervis Road, Bournemouth BH1 3DD 01202-291488
Fax 01202-290335
sB&B £35-£38.50

★★★*H* **East Anglia Hotel** 6 Poole Road, Bournemouth BH2 5QX 01202-765163

Fax 01202-752949
sB&B £37

★★★*HCR* **Langtry Manor Hotel** 26 Derby Road, Bournemouth BH1 3QB 01202-553887
Fax 01202-290115
sB&B £59.50

★★★**New Durley Dean Hotel** Westcliffe Road, Bournemouth BH2 5HE 01202-557711
Fax 01202-292815
sB&B £21.25-£48

★★★*H* **Queens Hotel** Meyrick Road, East Cliff, Bournemouth BH1 3DL 01202-554415
Fax 01202-294810
sB&B £39.50-£46.50

★★★**Trouville Hotel** 5 Priory Road, Bournemouth BH2 5DH 01202-552262
Fax 01202-293324
sB&B £39-£43.50

★★★**Wessex Hotel** 11 West Cliff Road, Bournemouth BH2 5EU 01202-551911
Fax 01202-297354
sB&B £42.95-£46.95

★★*HC* **Chinehurst Hotel** Studland Road, Westbourne, Bournemouth BH4 8JA 01202-764583 Fax 01202-762854
sB&B £22-£32

★★**Pinehurst Hotel** West Cliff Gardens, Bournemouth BH2 5HR 01202-556218
Fax 01202-551051
sB&B £19-£29

★★*H* **Winterbourne Hotel** Priory Road, West Cliff, Bournemouth BH2 5DJ 01202-296366
Fax 01202-780073
sB&B £25-£38

The Boltons Hotel *Highly Acclaimed* 9 Durley Chine Road South, Westcliff, Bournemouth BH2 5JT 01202-751517 Fax 01202-751629
sB&B £24-£26

Mae Mar Hotel 91-95 West Hill Road, Bournemouth BH2 5PQ 01202-553167
Fax 01202-311919
sB&B £16-£22.50

BOURTON-ON-THE-WATER *Gloucestershire*
★★★*H* **Old Manse Hotel** Victoria Street, Bourton-on-the-Water GL54 2BX 01451-820082
Fax 01451-810381
sB&B £33.50-£39.50

★★**Chester House Hotel** Bourton-on-the-Water GL54 2BU 01451-820286 Fax 01451-820471
sB&B £49

BRACKNELL *Berkshire*
★★★**Stirrups Country House Hotel** Maidens Green, Bracknell RG12 6LD 01344-882284
Fax 01344-882300

BRADFORD *West Yorkshire*
★★★**Novotel Motel** Merrydale Road, Bradford BD4 6SA 01274-683683 Fax 01274-651342
sB&B £47

★★★**Tong Village Hotel** The Pastures, Tong Lane, Bradford BD4 0RP 0113-2854646
Fax 0113-2853661

★★★**Victoria Hotel** Bridge Street, Bradford BD1 1JX 01274-728706 Fax 01274-736358
sB&B £68.50

BRADFORD-ON-AVON *Wiltshire*
★★★**Leigh Park Hotel** Leigh Road West, Bradford-on-Avon BA15 2RA 01225-864885
Fax 01225-862315
sB&B £69.50

BRAMHALL *Grt Manchester*
★★★*C* **Bramhall Moat House** Bramhall Lane South, Bramhall SK7 2EB 0161-4398116
Fax 0161-4408071
sB&B £72.50-£82.50

BRAMPTON *Cumbria*
★★**Tarn End House Hotel** Talkin Tarn, Brampton CA8 1LS 016977-2340
Fax 016977-2089
sB&B £29.50-£39.50

Oakwood House Hotel *Highly Acclaimed* Longtown Road, Brampton CA8 2AP 016977-2436
sB&B £26

BRANDS HATCH *Kent*
★★★★**Brands Hatch Thistle Hotel** Brands Hatch DA3 8PE 01474-854900 Fax 01474-853220
sB&B £49-£78.25

BRENTWOOD *Essex*
★★★★**Marygreen Manor Hotel** London Road, Brentwood CM14 4NR 01277-225252
Fax 01277-262809
sB&B £96.50

BRERETON *Staffordshire*
★★**Cedar Tree Hotel** Main Road, Rugeley, Brereton WS15 1DY 01889-584241
sB&B £23-£29

BRIDGNORTH *Shropshire*

❧ ★★★*HCR* **Old Vicarage Hotel** Worfield, Bridgnorth WV15 5JZ 01746-716497

Attractive, Edwardian vicarage in two acres of lovely grounds. Peace, quiet and seclusion guaranteed.
♿ parking, ramp (or level) access, adapted en-suite bathrooms, sB&B £65

⛺ **Stanmore Hall Touring Park** Stourbridge Road, Bridgnorth WV15 6DT 01743-232200 Fax 01743-353584
Car & Caravan £10.14

BRIDLINGTON *Humberside*

★★**Monarch Hotel** South Marine Drive, Bridlington YO15 3JJ 01262-674447 Fax 01262-604928
sB&B £28-£29.50

⛺ **South Cliff Caravan Park** Wilsthorpe, Bridlington YO15 3QN 01262-671051 Fax 01262-605639

BRIDPORT *Dorset*

Britmead House Hotel *Acclaimed* West Bay Road, Bridport DT6 4EG 01308-422941
sB&B £24-£33

⛺ **Binghams Farm Touring Caravan Park** Melplash, Bridport DT6 3TT 01308-488234
Car & Caravan £6.50

⛺ **Highlands End Farm Holiday Park** Eype, Bridport DT6 6AR 01308-422139 Fax 01308-425672

⛺ **West Bay Holiday Park** West Bay, Bridport DT6 4HB 01442-230300 Fax 01442-230368
Car & Caravan £7.50

BRIGG *Humberside*

★★★*C* **Briggate Lodge Inn** Ermine Street, Broughton, Brigg DN20 0NQ 01652-650770 Fax 01652-650495
sB&B £67

BRIGHTON *East Sussex*

★★★★★**Grand Hotel** Kings Road, Brighton BN1 2FW 01273-321188 Fax 01273-202694
sB&B £55-£130

★★★★★**The Brighton Thistle Hotel** Kings Road, Brighton BN1 2GS 01273-206700 Fax 01273-820692
♿ramp (or level) access, wheel chair Lift, adapted en-suite bathrooms, sB&B £124.75

★★★★**Bedford Hotel** Kings Road, Brighton BN1 2JF 01273-329744 Fax 01273-775877
A seafront hotel situated close to all of the town's many attractions. New outside terrace with sea view. Choice of several dining rooms and 24hr lounge available. Modern bedrooms.
♿ parking, ramp (or level) access, sB&B £92-£113

★★★**Brighton Oak Hotel** West Street, Brighton BN1 2RQ 01273-220033 Fax 01273-778000
sB&B £50.75-£86.75

BRISTOL *Avon*

★★★★**Aztec Hotel** Aztec West Business Park, Almondsbury, Bristol BS12 4TS 01454-201090 Fax 01454-201593
sB&B £83

★★★★**Grand Hotel** Broad Street, Bristol BS1 2EL 0117-9291645 Fax 0117-9227619
sB&B £93.75

★★★★**Holiday Inn Crowne Plaza** Victoria Street, Bristol BS1 6HY 0117-925-5010 Fax 0117-925-5040
sB&B £110

★★★★*CR* **Swallow Royal Hotel** College Green, Bristol BS1 5TA 0117-9255100 Fax 0117-9251515
The Swallow Royal Hotel stands majestically next to the Cathedral overlooking College Green. The Swallow Royal combines Victorian character and style with contemporary luxuries.
♿ parking, ramp (or level) access, wheel chair Lift, adapted en-suite bathrooms, sB&B £99.50

★★**The Glenroy Hotel** Victoria Square, Clifton, Bristol BS8 4EW2 0117-9739058
sB&B £46

Forte Crest Filton Road, Hambrook, Bristol BS16 1QX 0117-9564242 Fax 0117-9569735
sB&B £89

BROADWAY *Hereford & Worcester*

★★★★**Lygon Arms** Broadway WR12 7DU 01386-852255 Fax 01386-858611
sB&B £111.65

Leasow House Hotel *Highly Acclaimed* Laverton Meadows, Broadway WR12 7NA 01386-584526 Fax 01386-584596
sB&B £30-£40

Leedons Park Childswickham Road, Broadway WR12 7HB 01905-795999 Fax 01905-794012

BROCKENHURST *Hampshire*
★★★*C* **Balmer Lawn Hotel** Lyndhurst Road, Brockenhurst SO42 7ZB 01590-623116 Fax 01590-623864
& parking, ramp (or level) access, wheel chair Lift, adapted en-suite bathrooms, sB&B £55

★★★*HCR* **Carey's Manor Hotel** Lyndhurst Road, Brockenhurst SO42 7RH 01590-23551 Fax 01590-22799
sB&B £69

BROOK *Hampshire*
★★★*C* **Bell Inn** Nr Lyndhurst, Brook SO4 7JH 01703-812214 Fax 01703-813958
sB&B £39-£60

BUCKDEN *Cambridgeshire*
★★(Inn) *HC* **Lion Hotel** High Street, Nr Huntingdon, Buckden PE18 9XA 01480-810313 Fax 01480-811070
sB&B £55

BUCKINGHAM *Buckinghamshire*
★★★**Villiers Hotel** 3 Castle Street, Buckingham MK18 1BS 01208-822444 Fax 01208-822113
sB&B £45-£58

BUDE *Cornwall*
★★★**Hartland Hotel** Hartland Terrace, Bude EX23 8JY 01288-355661 Fax 01288-355664
sB&B £54-£60

★★**Maer Lodge Hotel** Maer Down Road, Bude EX23 8NG 01288-353306 Fax 01288-353406
sB&B £25-£28

★★**St Margarets Hotel** Killerton Road, Bude EX23 8EN 01288-352252 Fax 01288-355995
sB&B £25-£30

BURFORD *Oxfordshire*
★★(Inn) **Maytime** Asthall, Burford OX8 4HW 01993-822068
sB&B £43.60

BURNHAM MARKET *Norfolk*
★★★(Inn) **Hoste Arms** The Green, Burnham Market PE31 8JA 01328-738257 Fax 01328-730103
sB&B £48

BURNHAM-ON-SEA *Somerset*
Warren Farm Warren Road, Brean Sands, Burnham-on-Sea TA8 2RP. 01278-751227

BURNLEY *Lancashire*
★★**Friendly Stop Inn** Keirby Walk, Burnley BB11 2DH 01282-427611 Fax 01282-436370
sB&B £44.50-£63.50

BURSCOUGH *Lancashire*
★★★**Beaufort Hotel** High Lane, Burscough L40 7SN 01704-892655 Fax 01704-895135
sB&B £56

BURTON-UPON-TRENT *Staffordshire*
★★★(Inn) *C* **Queens Hotel** 2-5 Bridge Street, Burton-upon-Trent DE14 1SY 01283-564993 Fax 01283-517556
sB&B £59.50

★★★*R* **Riverside Inn** Riverside Drive, Branston, Burton-upon-Trent DE14 3EP 01283-511234 Fax 01283-511441
sB&B £31-£55

BURY ST EDMUNDS *Suffolk*
★★★**Butterfly Hotel** Symonds Road, Moreton Hall Estate, Bury St Edmunds IP32 7BW 01284-760884 Fax 01284-755476
sB&B £45.50-£55.50

BUXTON *Derbyshire*
★**Hartington Hotel** 18 Broad Walk, Buxton SK17 6JR 01298-22638
sB&B £20-£40

CALNE *Wiltshire*
Blackland Lakes Holiday & Leisure Centre Stockley Lane, Calne SN11 0NQ 01249-813672

CAMBRIDGE *Cambridgeshire*
★★★★**University Arms Hotel** Regent Street, Cambridge CB2 1AD 01223-351241 Fax 01223-315256
sB&B £96

★★**Quy Mill Hotel** Newmarket Road, Stow-cum-Quy, Cambridge CB5 9AG 01223-293383 Fax 01223-293770
sB&B £53.50

CANTERBURY *Kent*
★★★★**The County Hotel** High Street, Canterbury CT1 2RX 01227-766266 Fax 01277-451512
sB&B £77.50

South View Caravan Park Maypole Lane, Hoath, Canterbury CT3 4LL 01227-860280
Car & Caravan £8

CARLISLE *Cumbria*
★★★**County Hotel** 9 Botchergate, Carlisle CA1 1PQ 01228-31316 Fax 01228-515456
City centre hotel located in the heart of Carlisle. Recently refurbished (1990) to provide all modern conveniences, yet maintaining an atmosphere of

Georgian grandeur and opulence.
&ramp (or level) access, lift, wheel chair Lift, doors, adapted ground floor bedrooms, adapted en-suite bathrooms, adapted bathrooms, sB&B £33.95-£37

★★★**Cumbria Park Hotel** 32 Scotland Road, Carlisle CA3 9DG 01228-22887 Fax 01228-514796 sB&B £50

★★**Pinegrove** 262 London Road, Carlisle CA1 2QS 01228-24828 Fax 01228-810941 sB&B £42

★**Vallum House** Burgh Road, Carlisle CA2 7NB 01228-21860 sB&B £25

⚑ **Orton Grange Caravan Park** Wigton Road, Carlisle CA5 6LA 01228-710252 Car & Caravan £7.80

CASTLE CARY *Somerset*
★★**George Hotel** Market Place, Castle Cary BA7 7AH 01963-350761 Fax 01963-350035 sB&B £37-£45

CASTLE COMBE *Wiltshire*
ᐁ ★★★★**Manor House Hotel** Castle Combe SN14 7HR 01249-782206 Fax 01249-782159 sB&B £85-£105

CASTLE DONINGTON *Leicestershire*
★★★**Donington Thistle Hotel** East Midlands Inter. Airport, Castle Donington DE74 2SH 01332-850700 Fax 01332-850823 sB&B £93.95-£123.95

★★★**Priests House Hotel** Kings Mills, Castle Donington DE7 2RR 01332-810649 Fax 01332-811141 sB&B £72

Park Farmhouse Hotel *Acclaimed* Melbourne Road, Isley Walton, Castle Donington DE7 2RN 01332-862409 Fax 01332-862364 sB&B £36-£39.50

⚑ **Park Farmhouse Caravan Park** Melbourne Road, Isley Walton, Castle Donington DE74 2RN 01332-862409 Fax 01332-862364 Car & Caravan £4

CHADDESLEY CORBETT *Hereford & Worcester*
ᐁ ★★★*CR* **Brockencote Hall** Chaddesley Corbett DY10 4PY 01562-777871 Fax 01562-777872 sB&B £80

CHALFONT ST GILES *Buckinghamshire*
⚑ **Highclere Farm Country Touring Park** Newbarn Lane, Seer Green, Chalfont St Giles HP9 2QZ 01494-874505 Car & Caravan £9

CHARLECOTE *Warwickshire*
★★★*C* **Charlecote Pheasant Hotel** Charlecote CV35 9EW 01789-470333 Fax 01789-470222 sB&B £62-£76.50

CHEADLE *Staffordshire*
⚑ **Hales Hall Caravan & Camping Park** Oakamoor Road, Cheadle ST10 1BU 01538-753305 Fax 01782-202316 Car & Caravan £5

CHELMSFORD *Essex*
ᐁ ★★★★*HCR* **Pontlands Park Hotel** West Hanningfield Road, Great Baddow, Chelmsford CM2 8HR 01245-476444 Fax 01245-478393 sB&B £89.10-£105.60

Snows Oaklands Hotel *Highly Acclaimed* 240 Springfield Road, Chelmsford CM2 6BP 01245-352004 sB&B £30.36

CHELTENHAM *Gloucestershire*
★★★★**Cheltenham Park Hotel** Cirencester Road, Charlton Kings, Cheltenham GL53 8EA 01242-222021 Fax 01242-226935 sB&B £76-£86

★★★★**Queens Hotel** Promenade, Cheltenham GL50 1NN 01242-514724 Fax 01242-224145 sB&B £44-£59

ᐁ ★★★**The Greenway** Shurdington, Cheltenham GL51 5UG 01242-862352 Fax 01242-862780 sB&B £77.50

★★★**The Prestbury House Hotel** The Burbage, Prestbury, Cheltenham GL52 3DN 01242-529533 Fax 01242-227076 sB&B £50-£65

★★*C* **Charlton Kings Hotel** London Road, Charlton Kings, Cheltenham GL52 6UU 01242-231061 Fax 01242-241900 sB&B £39-£64

Hilden Lodge Hotel *Highly Acclaimed* 271 London Road, Charlton Kings, Cheltenham GL52 6YL 01242-583242 Fax 01242-263511 sB&B £30

⚑ **Folly Farm** Notgrove, Cheltenham GL54 3BY 01242-820285

CHESTER *Cheshire*
★★★★*C* **Birches Hotel and Carden Park** Chester 01244-731000 Fax 01829-250539 sB&B £85

★★★★**Chester Moat House International** Trinity Street, Chester CH1 2BD 01244-322330 Fax 01244-316118

Luxury hotel, located in the heart of this ancient city. Disabled travellers are warmly welcomed. Good access provided.
&. parking, lift, wheel chair Lift, good access to restaurant, doors, adapted en-suite bathrooms, adapted bathrooms, sB&B £99

🍴 ★★★★HCR **Crabwall Manor Hotel** Parkgate Road, Mollington, Chester CH1 6NE 01244-851666 Fax 01244-851100
&. parking, ramp (or level) access, lift, doors, adapted ground floor bedrooms, adapted en-suite bathrooms, adapted bathrooms, sB&B £98.50

★★★C **Hoole Hall Hotel** Warrington Road, Hoole Village, Chester CH2 3PD 01244-350011 Fax 01244-320251
sB&B £61

★★★**Rowton Hall Hotel** Whitchurch Road, Rowton, Chester CH3 6AF 01244-335262 Fax 01244-335464
sB&B £72

★★**Brookside Hotel** Brook Lane, Chester CH2 2AN 01244-381943 Fax 01244-379701
sB&B £30

★★**Dene Hotel** 95 Hoole Road, Chester CH2 3ND 01244-321165 Fax 01244-350277
sB&B £37

★★HC **Green Bough Hotel** 60 Hoole Road, Hoole, Chester CH2 3NL 01244-326241 Fax 01244-326265
sB&B £38-£43

Eaton Hotel 29 City Road, Chester CH1 3AE 01244-320840 Fax 01244-320850
sB&B £28

CHICHESTER *West Sussex*
★★★**Chichester Resort** Westhampnett, Chichester PO19 4UL 01243-782371 Fax 01243-786351
sB&B £63.50-£73.25

🅰🚐 **Red House Farm** Earnley, Chichester PO20 7JG 01243-512959

CHIPPING CAMPDEN *Gloucestershire*
★★(Inn) **Noel Arms Hotel** High Street, Chipping Campden GL55 6AT 01386-840317 Fax 01386-841136
sB&B £58

CHIPPING NORTON *Oxfordshire*
★★**Crown & Cushion Hotel** 23 High Street, Chipping Norton OX7 5AD 01608-642533 Fax 01608-642926
sB&B £35-£39

CHISLEHAMPTON *Oxfordshire*
★★(Inn) **Coach and Horses Inn** Stadhampton Road, Chislehampton, Oxford OX44 7UX 01865-890255 Fax 01865-891995
sB&B £45

CHOLDERTON *Wiltshire*
Fayre Deal Motel Parkhouse Corner, Cholderton SP4 0EG 01980-629542 Fax 01980-629542
sB&B £31.75

CHOLLERFORD *Northumberland*
★★★**George Hotel** Chollerford NE46 4EW 01434-681611 Fax 01434-681727
&. parking, doors, adapted ground floor bedrooms, adapted en-suite bathrooms, adapted bathrooms, sB&B £87.50

CHRISTCHURCH *Dorset*
★★HC **Fisherman's Haunt Hotel** Winkton, Christchurch BH23 7AS 01202-484071 Fax 01202-478883
sB&B £36

🅰🚐 **Heathfield Caravan & Camp Site** Avon Causeway, Hurn., Christchurch 01202-485208
Car & Caravan £6

🅰🚐 **Mount Pleasant Touring Park** Matchams Lane, Hurn, Christchurch BH23 6AW 01202-475474 Fax 01202-483878
Car & Caravan £10

CHUDLEIGH *Devon*
🅰🚐 **Holmans Wood Tourist Park** Chudleigh TQ13 0DZ 01626-853785
Car & Caravan £5.95

CHURCH OAKLEY *Hampshire*
★★★(Inn) **Beach Arms** Nr Basingstoke, Church Oakley 01256-780210 Fax 01256-780557
sB&B £55

CLAYTON-LE-WOODS *Lancashire*
★★★CR **The Pines Hotel** Preston Road, Clayton-le-Woods PR6 7ED 01772-38551 Fax 01772-38551
sB&B £50

CLEVEDON *Avon*
(Inn) **Salthouse** Salthouse Road, Clevedon BS21 7TY 01275-871482
sB&B £30

COALVILLE *Leicestershire*
★★★**Hermitage Park Hotel** Whitwick Road, Coalville LE67 3FA 01530-814814 Fax 01530-814202
sB&B £59.50

PLACES TO STAY

COLCHESTER *Essex*
★★★**Butterfly Hotel** A12/A120 Junction, Old Ipswich Road, Colchester CO7 7QY 01206-230900 Fax 01206-231095
sB&B £55.50

★★★*C* **Marks Tey Hotel** London Road, Marks Tey, Colchester CO6 1DU 01206-210001 Fax 01206-212167
sB&B £60

COLEFORD *Gloucestershire*
★★(Inn) **Orepool Inn** St Briavels Road, Sling, Coleford GL16 8LH 01594-833277 Fax 01594-833785

& parking, ramp (or level) access, adapted en-suite bathrooms, sB&B £34.75

COLYTON *Devon*
★★(Inn) **White Cottage Hotel** Dolphin Street, Colyton EX13 6NA 01297-552401 Fax 01297-553897
sB&B £34.90

▲♥ **Leacroft Touring Park** Colyton 01297-552823
Car & Caravan £7.50

CONISTON *Cumbria*
★★(Inn) **Black Bull** Yewdale Road, Coniston LA21 8DU 0153-94-41335 Fax 0153-94-41168
sB&B £33

COODEN BEACH *East Sussex*
★★★**Cooden Resort Hotel** Cooden Sea Road, Cooden Beach TN39 4TT 01424-842281 Fax 01424-846142
sB&B £55

CORNHILL-ON-TWEED *Northumberland*
★★**Collingwood Arms Hotel** Cornhill-on-Tweed TD12 4UH 01890-882424
sB&B £35

CORSHAM *Wiltshire*
★★(Inn) *CR* **Methuen Arms Hotel** High Street, Corsham SN13 0HB 01249-714867

Fax 01249-712004
sB&B £47

COULSDON *Surrey*
♠ ★★★★*H* **Coulsdon Manor Hotel** Coulsdon Court Road, Nr Croydon, Coulsdon CR5 2LL 0181-668-0414 Fax 0181-668-3118
sB&B £76

COVENTRY *West Midlands*
★★★★**De Vere Hotel** Cathedral Square, Coventry CV1 5RP 01203-633733 Fax 01203-225299
sB&B £85-£95

★★★*HC* **Brooklands Grange Hotel** Holyhead Road, Coventry CV5 8HX 01203-601601 Fax 01203-601277
sB&B £45-£75

★★★**Coventry Hill Hotel** Rye Hill, Allesley, Coventry CV5 9PH 01203-402151 Fax 01203-402235
sB&B £63-£73

Campanile Abbey Road, Whitley, Coventry CV3 4BJ 01203-639922
sB&B £40

Falcon 13-19, Manor Road, Coventry CV1 2LH 01203-258615 Fax 01203-520680
sB&B £28.50-£35

CROOKLANDS *Cumbria*
▲♥ **Waters Edge Caravan Park** Crooklands LA7 7NN 015395-67708 Fax 015395-67610
Car & Caravan £12.75

CROWTHORNE *Berkshire*
★★★**Waterloo Hotel** Duke's Ride, Crowthorne RG11 7NW 01344-777711 Fax 01344-778913
sB&B £73.50

CROYDON *Surrey*
★★★★**Croydon Park Hotel** 7 Altyre Road, Croydon CR9 5AA 0181-680-9200 Fax 0181-760-0426
A modern, executive hotel, in the centre of Croydon's thriving business community.
& parking, ramp (or level) access, wheel chair Lift, adapted en-suite bathrooms, sB&B £87

DARLINGTON *Co Durham*
★★★**Swallow Kings Head Hotel** Priestgate, Darlington DL1 1NW 01325-380222 Fax 01325-382006
sB&B £75

▲♥ **Winston Caravan Park** Winston, Darlington DL2 3RH 01325-730228 Fax 01325-730228
Car & Caravan £6.50

DARTMOUTH *Devon*
👥🚢 **Little Cotton Caravan Park** Dartmouth TQ6 0LB 01803-832558
Car & Caravan £5.75

👥🚢 **Woodland Leisure Park** Blackawton, Dartmouth TQ9 7DQ 01803-712598
Fax 01803-712680

DARWEN *Lancashire*
★★★**Whitehall Hotel** Springbank, Whitehall, Darwen BB3 2JU 01254-701595 Fax 01254-773426 sB&B £49.50

DAWLISH *Devon*
★★★**Langstone Cliff Hotel** Dawlish EX7 0NA 01626-865155 Fax 01626-867166
The hotel is on two floors only with the older part dating from c.1750, but with modern extensions. Situated in twenty wooded acres, overlooking the sea.
♿ parking, ramp (or level) access, wheel chair Lift, sB&B £35-£40

West Hatch Hotel *Acclaimed* 34 West Cliff, Dawlish EX7 9DN 01626-864211
sB&B £28-£38

👥🚢 **Cofton Country Holiday Park** Starcross, Dawlish EX6 8RP 01626-890111
Fax 01626-891572
Car & Caravan £5.80

DERBY *Derbyshire*
★★★*HC* **Midland Hotel** Midland Road, Derby DE1 2SQ 01332-345894 Fax 01332-293522
sB&B £74.50

★★*HCR* **Hotel La Gondola** 220 Osmaston Road, Derby DE23 8JX 01332-332895 Fax 01332-384512
sB&B £47.50

★★*R* **Kedleston Hotel** Kedleston Road, Derby DE22 5JD 01332-559202 Fax 01332-558822
sB&B £42.50

DONCASTER *South Yorkshire*
★★★**Danum Swallow Hotel** High Street, Doncaster DN1 1DN 01302-342561
Fax 01302-329034
sB&B £78

★★★**Doncaster Moat House** Warmsworth, Doncaster DN4 9UX 01302-310331
Fax 01302-310197
sB&B £84

♿ ★★★**Mount Pleasant Hotel** Great North Road (A638), Rossington, Doncaster DN11 0HP 01302-868219 Fax 01302-865130
sB&B £45

DORCHESTER *Dorset*
Casterbridge Hotel *Highly Acclaimed* 49, High East Street,, Dorchester DT1 1HU 01305-264043

Fax 01305-260884
sB&B £30-£36

DORKING *Surrey*
(Inn) The Pilgrim Station Road, Dorking RH4 1HF 01306-889951
sB&B £22

DRIFFIELD, GREAT *Humberside*
★★★*HCR* **The Bell** Market Place, Driffield, Great YO25 7AP 01377-46661 Fax 01377-43228
sB&B £65

DUDLEY *West Midlands*
★★★★**Copthorne Merry Hill** Level Street, Brierley Hill, Dudley DY5 1UR 01384-482882
Fax 01384-482773
sB&B £101.75

★★**Station Hotel** Castle Hill, Dudley DY1 4RA 01384-253418 Fax 01384-457503
sB&B £39.50

DULVERTON *Somerset*
👥🚢 **Exmoor House Caravan Club Site** Dulverton TA22 9HL 01398-23268

DUNSTER *Somerset*
★★*HC* **Exmoor House Hotel** West Street, Dunster TA24 6SN 01643-821268
sB&B £28.50

DUNWICH *Norfolk*
👥🚢 **Cliff House** Minsmere Road, Dunwich IP17 3DQ 0172-873-282
Car & Caravan £7

EAGLESCLIFFE *Cleveland*
★★★**Parkmore Hotel** 636 Yarm Road, Eaglescliffe TS16 0DH 01642-786815
Fax 01642-790485
Victorian building, extensively modernised and extended into a hotel providing excellent facilities.
♿ramp (or level) access, adapted en-suite bathrooms, sB&B £55-£59

EAST DEREHAM *Norfolk*
★★**(Inn) Kings Head** Norwich Street, East Dereham NR19 1AD 01362-693842
sB&B £36

EASTBOURNE *East Sussex*
★★★★★**Grand Hotel** King Edwards Parade, Eastbourne BN21 4EQ 01323-412345
Fax 01323-412233
sB&B £91.50

★★★**Mansion Hotel** Grand Parade, Eastbourne BN21 3YS 01323-727411 Fax 01323-720665
sB&B £58.75

★★*H* **Congress Hotel** 31 Carlisle Road, Eastbourne BN21 4JS 01323-732118

Fax 01323-720016
sB&B £28-£31

★★**Farrar's Hotel** 3/5 Wilmington Gardens,
Eastbourne BN21 4JN 01323-723737
Fax 01323-732902
sB&B £20-£30

★★**Langham Hotel** Royal Parade, Eastbourne
BN22 7AH 01323-731451 Fax 01323-646623
sB&B £19.95-£24.95

Flamingo Hotel *Acclaimed* 20 Enys Road,
Eastbourne BN21 2DN 01323-721654
sB&B £21-£22.50

EBCHESTER *Co Durham*

★★★**Raven Hotel** Broomhill, Ebchester DH8
6RY 01207-560367 Fax 01207-560262
sB&B £52

EGHAM *Surrey*

★★★★*HCR* **Runnymede Hotel** Windsor Road,
Egham TW20 0AG 01784-436171
Fax 01784-436340
sB&B £123.95

EGREMONT *Cumbria*

★★★**Blackbeck Bridge Inn** Beckermet,
Egremont CA22 2NY 01946-841661
Fax 01946-841007
sB&B £49.50

ELLESMERE *Shropshire*

▲🚐 **Fernwood Caravan Park** Lyneal, Ellesmere
SY12 0GF 01948-75221
Car & Caravan £9

ELY *Cambridgeshire*

♽ ★**Nyton Hotel** 7 Barton Road, Ely CB7 4HZ
01353-662459
sB&B £34

EMSWORTH *Hampshire*

★★★*C* **Brookfield Hotel** Havant Road,
Emsworth PO10 7LF 01243-373555
Fax 01243-376342
sB&B £49

EPWORTH *Humberside*

★★(Inn) **Red Lion Hotel** Market Place, Epworth
DN9 1EU 01427-872208 Fax 01427-874330
sB&B £39.50

ESHER *Surrey*

Lakewood House *Acclaimed* Portsmouth Road,
Esher KT10 9JH 01932-867142 Fax 01923-867142
sB&B £35.15

EVESHAM *Worcestershire*

★★★**Northwick Arms Hotel** Waterside, Evesham
WR11 6BT 01386-40322 Fax 01386-41070

*Hotel in a picturesque waterside setting on the edge of the
Cotswolds, close to Stratford, Warwick and Cheltenham.*
♿ *parking, doors, adapted ground floor
bedrooms, adapted en-suite bathrooms.*
sB&B £50

EXETER *Devon*

★★★**Buckerell Lodge Hotel** Topsham Road,
Exeter EX2 4SQ 01392-52451 Fax 01392-412114
sB&B £42.95

★★★**Granada Lodge** Moor Lane, Sandygate,
Exeter EX2 4AR 01392-74044 Fax 01392-410406
sB&B £51

★★*HC* **St Andrews Hotel** 28 Alphington Road,
Exeter EX2 8HN 01392-76784 Fax 01392-50249
sB&B £39

▲🚐 **Webbers Farm Caravan Park** Castle Lane,
Woodbury, Exeter EX5 1EA 01395-232276
Fax 01395-232276
Car & Caravan £7.45

FAKENHAM *Norfolk*

★★(Inn) **Crown Hotel** Market Place, Fakenham
NR21 9BP 01328-851418
sB&B £42

▲🚐 **Fakenham Racecourse** Fakenham NR21
7NY 01328-862388
Car & Caravan £9

FALMOUTH *Cornwall*

★★★**Green Lawns Hotel** Western Terrace,
Falmouth TR11 4QJ 01326-312734
Fax 01326-211427
sB&B £44

FAREHAM *Hampshire*

★★★★*HC* **Solent Hotel** Solent Business Park,
Whiteley, Fareham PO15 7AJ 01489-880000
Fax 01489-880007
sB&B £82

Avenue House Hotel *Acclaimed* 22 The Avenue, Fareham PO14 1NS 01329-232175 Fax 01329-232196
Former private residence converted in 1988 and retaining the atmosphere of a family home. Surrounded by mature gardens. Located close to the town centre and 400 yards west of Fareham railway station.
&ramp (or level) access, adapted en-suite bathrooms, sB&B £39.50-£45

FELIXSTOWE *Suffolk*
★★**Marlborough Hotel** Sea Road, Felixstowe IP11 8BJ 01394-285621 Fax 01394-670724 sB&B £44.50-£47

⚠☎ **Suffolk Sands Caravan Park** Carr Road, Felixstowe IP11 8TS 01394-273314
Car & Caravan £6.20

FILEY *North Yorkshire*
⚠☎ **Blue Dolphin Holiday Centre** Gristhorpe Bay, Filey YO14 9PU 01442-230300 Fax 01442-230368
Car & Caravan £7.20

FORDINGBRIDGE *Hampshire*
⚠☎ **New Forest Country Holidays** Sandy Balls Estate, Godshill, Fordingbridge SP6 2JZ 01425-653042 Fax 01425-653067

FOWEY *Cornwall*
★★★**Fowey Hotel** The Esplanade, Fowey PL23 1HX 01726-832551 Fax 01726-832125 sB&B £44

Carnethic House Hotel *Highly Acclaimed* Lambs Barn, Fowey PL23 1HQ 01726-833336
sB&B £30-£40

FOWNHOPE *Hereford & Worcester*
★★(Inn) *C* **Green Man Inn Hotel** Fownhope HR1 4PE 01432-860243 Fax 01432-860207 sB&B £30-£31

FRIMLEY *Surrey*
★★★**One Oak Toby** 114 Portsmouth Road, Frimley GU15 1HS 01276-691939 Fax 01276-676088
A distinctive new hotel built behind the traditional pub with elegant and comfortable furnishings.
&ramp (or level) access, adapted en-suite bathrooms, sB&B £69.25

GARBOLDISHAM *Norfolk*
Ingleneuk Lodge *Highly Acclaimed* Hopton Road, Diss, Garboldisham IP22 2RQ 0195-381-541 sB&B £31.50

GARSTANG *Lancashire*
⚠☎ **Claylands Caravan Park** Claylands Farm, Cabus, Garstang PR3 1AJ 01524-791242
Car & Caravan £10.20

GATESHEAD *Tyne & Wear*
★★★**Swallow Hotel** High West Street, Gateshead NE8 1PE 01914-771105 Fax 01914-787214 sB&B £78

GATWICK AIRPORT *West Sussex*
Forte Crest North Terminal, Crawley RH6 0PH 01293-567070 Fax 01293-567739
sB&B £108.50

★★★**Gatwick Moat House** Longbridge Roundabout, Horley RH6 0AB 01293-785599 Fax 01293-785991
sB&B £54.50-£81.95

⚠ ★★*HC* **Langshott Manor** Langshott, Horley RH6 9LN 01293-786680 Fax 01293-783905 sB&B £86

GISBURN *Lancashire*
★★★**Stirk House Hotel** Gisburn BB7 4LJ 01200-445581 Fax 01200-445744 sB&B £45

GLASTONBURY *Somerset*
☎ **Cradlebridge Farm** Glastonbury BA16 9SD 01458-831827
sB&B £25

⚠☎ **Old Oaks Touring Park** Wick Farm, Wick, Glastonbury BA6 0JS 01458-831437
Car & Caravan £7

GLOSSOP *Derbyshire*
George Hotel 34 Norfolk Street, Glossop SK13 9QU 01457-855449 Fax 01457-857033 sB&B £25

GOATHLAND *North Yorkshire*
Heatherdene *Acclaimed* Goathland YO22 5AN 01947-86334
sB&B £28

GOODWOOD *West Sussex*
★★★**Goodwood Park Hotel** Goodwood PO18 0QB 01243-775537 Fax 01243-533802 sB&B £64-£84

GORLESTON-ON-SEA *Norfolk*
Squirrel's Nest *Acclaimed* 71 Avondale Road, Gorleston-on-Sea NR31 6DJ 01493-662746 sB&B £15-£27

GRANGE-OVER-SANDS *Cumbria*
★★★*H* **Netherwood Hotel** Lindale Road, Grange-over-Sands LA11 6ET 01539-532552 Fax 01539-534121
sB&B £42.75-£51.50

PLACES TO STAY

GRANTHAM *Lincolnshire*
★★★★*HC* **Belton Woods Hotel & Club** Belton, Grantham NG32 2LN 01476-593200
Fax 01476-74547
sB&B £90

★★★*C* **Swallow Hotel** Swingbridge Road, Grantham NG31 7XT 01476-593000
Fax 01476-592592
Newly built, two storey building with traditional furnishing and rich fabrics. Courtyard-style hotel with a welcoming atmosphere.
& parking, ramp (or level) access, doors, adapted ground floor bedrooms, adapted en-suite bathrooms, adapted bathrooms, sB&B £85

★★*HC* **Kings Hotel** North Parade, Grantham NG31 8AU 01476-590800 Fax 01476-590800
sB&B £45

GRASMERE *Cumbria*
★★★★*HR* **Wordsworth** Grasmere LA22 9SW 0153-94-35592 Fax 0153-94-35765
sB&B £58.50

★★★*HR* **Gold Rill Hotel** Red Bank Road, Grasmere LA22 9PU 0153-94-35486
Fax 0153-94-35486
sB&B £25-£40

GREAT YARMOUTH *Norfolk*
★★**Burlington Hotel** North Drive, Great Yarmouth NR30 1EG 01493-844568
Fax 01493-331848
& parking, ramp (or level) access, wheel chair lift, doors, adapted ground floor bedrooms, adapted en-suite bathrooms, sB&B £35-£55

★★**Palm Court Hotel** North Drive, Great Yarmouth NR30 1EF 01493-844568
Fax 01493-331848
sB&B £39-£55

★★**Royal Hotel** Marine Parade, Great Yarmouth NR30 3AE 01493-844215
sB&B £22

▲♥ **Willowcroft Camping and Caravan Park** Staithe Road, Repps-with-Bastwick, Great Yarmouth NR29 5JU 01692-670380
Car & Caravan £7

▲♥ **Clippesby Holidays** Clippesby, Great Yarmouth NR29 3BJ 01493-369367
Car & Caravan £11

▲♥ **Liffens Holiday Park** Burgh Castle, Great Yarmouth NR31 9QB 01493-780357
Car & Caravan £9

▲♥ **Seashore Holiday Super Centre** North Denes, Great Yarmouth NR30 4HG 01493-851131
Fax 01442-230368
Car & Caravan £6.50

HADLEY WOOD *Gtr London*
♿ ★★★★*CR* **West Lodge Park Hotel** Cockfosters Road, Hadley Wood EN4 0PY 0181-4408311
Fax 0181-4493698
sB&B £73.25-£104.75

HAILSHAM *East Sussex*
▲♥ **The Old Mill Caravan Park** Chalvington Road, Golden Cross, Hailsham BN27 3SS 01825-872532
Car & Caravan £5

HALIFAX *West Yorkshire*
♿ ★★★*HCR* **Holdsworth House Hotel** Holdsworth, Halifax HX2 9TG 01422-240024
Fax 01422-245174
& ramp (or level) access, adapted en-suite bathrooms, sB&B £69.70

HALLAND *East Sussex*
★★★**Halland Forge Hotel** Halland BN8 6PW 01825-840456 Fax 01825-840773
sB&B £44

HANDFORTH *Cheshire*
★★★★**Belfry Hotel** Stanley Road, Handforth SK9 3LD 0161-4370511 Fax 0161-4990597
sB&B £79

HARROGATE *North Yorkshire*
★★★★*R* **Boars Head Hotel** Ripley, Harrogate HG3 3AY 01423-771888 Fax 01423-771509
sB&B £70

★★★★**Majestic** Ripon Road, Harrogate HG1 2HU 01423-568972 Fax 01423-502283
sB&B £78.50-£98.50

★★★★**Moat House International** Kings Road, Harrogate HG1 1XX 01423-500000
Fax 01423-524435
sB&B £59-£109

★★★*H* **Grants Hotel** 3-13 Swan Road, Harrogate HG1 2SS 01423-560666 Fax 01423-502550
sB&B £49.95-£87.50

Alexa House Hotel *Acclaimed* 26 Ripon Road, Harrogate HG2 2JJ 01423-501988
Fax 01423-504086
sB&B £30-£35

▲♥ **High Moor Farm Park** Skipton Road, Harrogate HG3 2LZ 01563-637
Car & Caravan £8

HARWELL *Oxfordshire*
★★★**Kingswell Hotel** Reading Road, Harwell OX11 0LZ 01235-833043 Fax 01235-833193
sB&B £72.50

HATFIELD HEATH *Hertfordshire*
♿ ★★★★*C* **Down Hall Country House Hotel** Nr. Bishop's Stortford, Hatfield Heath CM22 7AS 01279-731441 Fax 01279-730416
sB&B £89.25

HAVANT *Hampshire*
★★★**Bear Hotel** East Street, Havant PO9 1AA 01705-486501 Fax 01705-470551
The Bear hotel is a genuine coaching inn, recently redecorated in a traditional style. Providing a warm and friendly atmosphere, as well as high standards of cuisine and comfort.
♿ramp (or level) access, sB&B £50

HAYLING ISLAND *Hampshire*
▲⊕ **Fishery Creek Camping & Caravan Park** Fishery Lane, Hayling Island PO11 9NR 01705-462164
Car & Caravan £5.50

HEADLAM *Co Durham*
♿ ★★★**Headlam Hall Hotel** Gainford, Darlington, Headlam DL2 3HA 01325-730238 Fax 01325-730790
sB&B £55-£60

HELLIDON *Northamptonshire*
★★★★**Hellidon Lakes Country Club** nr Daventry, Hellidon NN11 6LN 01327-62550 Fax 01327-62159
sB&B £85

HELMSLEY *North Yorkshire*
▲⊕ **Foxholme Touring Caravan Park** Harome, Helmsley YO6 5JG 01439-770316
Car & Caravan £6.50

▲⊕ **Golden Square Caravan Park** Oswaldkirk, Helmsley YO6 5YQ 01439-788659
Car & Caravan £5.20

HENFIELD *West Sussex*
R **Tottington Manor Hotel** *Highly Acclaimed* Edburton, Henfield BN5 9LJ 01903-815757 Fax 01903-879131
sB&B £35-£50

HEREFORD *Hereford & Worcester*
★★★**Hereford Moat House** Belmont Road, Hereford & Worcester HR2 7BP 01432-354301 Fax 01432-275114
sB&B £63

HINCKLEY *Leicestershire*
★★★**Hinckley Island Hotel** A5 Watling Street, Hinckley LE10 3JA 01455-631122 Fax 01455-634536
sB&B £74

★★★**Sketchley Grange Hotel** Sketchley Grange, Burbage, Hinckley LE10 3HU 01455-251133

Fax 01455-631384
sB&B £73

HOCKLEY HEATH *Warwickshire*
♿ ★★★*HCR* **Nuthurst Grange** Hockley Heath B94 5NL 01564-783972 Fax 01564-783919
sB&B £89

HOLMES CHAPEL *Cheshire*
★★★**Old Vicarage Hotel** Knutsford Road, Cranage, Holmes Chapel CW4 8EF 01477-532041 Fax 01477-535728
Authentic 17th century, Grade II listed, building with modern bedroom extensions. Situated on the banks of the River Dane.
♿ parking, ramp (or level) access, adapted ground floor bedrooms, adapted en-suite bathrooms, adapted bathrooms, sB&B £65

HOLMFIRTH *West Yorkshire*
White Horse Inn Jackson Bridge, Holmfirth HD7 7HF 01484-683940
sB&B £24

HOLT *Norfolk*
★★**Pheasant Hotel** The Coast Road, Kelling, Holt NR25 7EG 01263-588382 Fax 01263-588101
sB&B £36

HOLYWELL BAY *Cornwall*
▲⊕ **Trevornick Holiday Park** Holywell Bay TR8 5PW 01637-830531
Car & Caravan £.70

HONILEY *Warwickshire*
★★★**Honiley Court Hotel** Honiley Road (A4117), Honiley CV8 1NP 01926-484234 Fax 01926-484474
sB&B £66.50

HORNS CROSS *Devon*
♿ ★★★*R* **Foxdown Manor Hotel** Bideford, Horns Cross EX39 5PJ 01237-451325 Fax 01237-451525
sB&B £55

HORTON *Dorset*
Northill House Hotel *Highly Acclaimed*
Wimborne, Horton BH21 7HL 01258-840407
A mid-19th century farmhouse with all modern conveniences. Winner of the 1991 "Peter Fry Award" for exellent service to the disabled.
♿ parking, ramp (or level) access, doors, adapted ground floor bedrooms, adapted en-suite bathrooms, adapted bathrooms, sB&B £37

HOUNSLOW *Gtr London (Middx)*
Heathrow *Acclaimed* 17-19 Haslemere Avenue, Hounslow TW5 9UT 0181-3843333 Fax 0181-3843321
sB&B £25-£30

Crompton 49 Lampton Road, Hounslow TW3 1JG 0181-5707090 Fax 0181-5771975
sB&B £30

HULL *Humberside*
★★★**Royal Hotel** Ferensway, Hull HU1 3UF 01482-325087 Fax 01482-323172
sB&B £67.90-£81.10

HUNGERFORD *Berkshire*
Marshgate Cottage *Acclaimed* Marsh Lane, Hungerford RG17 0QX 01488-682307 Fax 01488-685475
sB&B £35.40

HUNSTANTON *Norfolk*
★★**Caley Hall Motel** Old Hunstanton, Hunstanton PE36 6HH 01485-533486 Fax 01485-533348

sB&B £29-£32

HUNTINGDON *Cambridgeshire*
▲⚐ **Park Lane Touring Park** Park Lane, Godmanchester, Huntingdon PE18 8AF 01480-453740

ILFORD *Essex*
Cranbrook Hotel *Acclaimed* 24 Coventry Road, Ilford IG1 4QR 0181-554-6544 Fax 0181-518-1463
sB&B £25.02

ILFRACOMBE *Devon*
▲⚐ **Hidden Valley Touring & Camping Park**
West Down, Ilfracombe EX34 8NU 01271-813837
Car & Caravan £8

INGATESTONE *Essex*
★★★**Heybridge** Roman Road, Ingatestone CM4 9AB 01277-355355 Fax 01277-353288
A Tudor style building, part of which dates back to 1494, set in the heart of Essex countryside in the pretty village of Ingatestone. Modern bedroom accommodation and excellent conference facilities.
♿ramp (or level) access, adapted en-suite bathrooms, sB&B £85.50

INSTOW *Devon*
Anchorage Hotel *Acclaimed* The Quay, Bideford, Instow EX39 4HX 01271-860655
sB&B £21-£23

IPSWICH *Suffolk*
★★★**Ipswich Moat House** London Road, Ipswich IP8 3JD 01473-730444 Fax 01473-730801
sB&B £63

★★★**Novotel** Greyfriars Road, Ipswich IP1 1UP 01473-232400 Fax 01473-232414
sB&B £57

★★★*C* **Suffolk Grange Hotel** The Havens, Ransomes Europark, Ipswich IP3 9SJ 01473-272244 Fax 01473-272484
♿ parking, ramp (or level) access, wheel chair Lift, adapted en-suite bathrooms, sB&B £66.50

▲⚐ **Low House Touring Caravan Centre**
Bucklesham Road, Foxhall, Ipswich IP10 OAU 01473-659437

ISLE OF WIGHT
Cygnet Hotel *Acclaimed* 58 Carter Street, Sandown PO36 8DQ 01983-402930 Fax 01983-405112
sB&B £21-£26

Norton Lodge Hotel 22 Victoria Road, Sandown PO36 8AL 01983-402423
sB&B £13.50-£17

★★*H* **Keats Green Hotel** 3 Queens Road, Shanklin PO37 6AN 01983-862742 Fax 01983-868868
sB&B £30

★★**Luccombe Hall Hotel** Luccombe Road, Shanklin PO37 6RL 01983-862719 Fax 01983-863082
sB&B £34-£37

Carlton Hotel East Cliff Promenade, Shanklin PO37 6AY 01983-862517
sB&B £22-£25

▲⚏ **Lower Hyde Character Holiday Village** Shanklin PO37 7LL 01442-230300
Car & Caravan £7.50

★★★*H* **Country Garden Hotel** Church Hill, Totland Bay PO39 0ET 01983-754521
sB&B £36

IVYBRIDGE *Devon*
★★**(Inn) Sportsman Inn** Exeter Road, Ivybridge PL21 0BQ 01752-892280 Fax 01752-690714
sB&B £35

KEGWORTH *Leicestershire*
★★★**Yew Lodge Hotel** Kegworth DE7 2DF 01509-672518 Fax 01509-674730
sB&B £30

Kegworth Hotel Packington Hill, Kegworth DE7 2DF 01509-672427 Fax 01509-674344
sB&B £46.95

KENDAL *Cumbria*
★★**Garden House Hotel** Fowl-ing Lane, Kendal LA9 6PH 01539-731131 Fax 01539-740064
sB&B £45-£50

KENILWORTH *Warwickshire*
★★★**De Montfort Hotel** The Square, Kenilworth CV8 1ED 01926-55944
Fax 01926-57830
sB&B £50-£80

KESSINGLAND *Suffolk*
▲⚏ **Heathland Beach Caravan Park** London Road, Kessingland NR33 7PJ 01502-740377
Car & Caravan £8.20

KESWICK *Cumbria*
★★★**Derwentwater Hotel** Portinscale, Keswick CA12 5RE 017687-72538 Fax 017687-71002
sB&B £60

★★**Chaucer House Hotel** Derwentwater Place, Keswick CA12 4DR 017687-72318
Fax 017687-75551
sB&B £29-£34.70

KETTERING *Northamptonshire*
★★★★**Kettering Park Hotel** *Highly Acclaimed* Kettering Parkway, Kettering NN15 6XT 01536-416666 Fax 01536-416171
sB&B £90

KETTLEWELL *North Yorkshire*
Langcliffe Country Guest House *Highly Acclaimed* Kettlewell BD23 5RJ 01756-760321
sB&B £30-£35

KEXBY *North Yorkshire*
⚏ **Ivy House Farm** Kexby, York YO4 5LQ 01904-489368
sB&B £55

KIDDERMINSTER *Hereford & Worcester*
Cedars Hotel *Highly Acclaimed* Mason Road, Kidderminster DY11 6AL 01562-515595
Fax 01562-753110
sB&B £49.80

KIDLINGTON *Oxfordshire*
Bowood House *Highly Acclaimed* 238 Oxford Road, Kidlington OX5 1EB 01865-842288
Fax 01865-841858
sB&B £43

KIELDER *Northumberland*
▲⚏ **Kielder Campsite** Hexham, Kielder NE48 1EL 01434-250291
Car & Caravan £9

KING'S LYNN *Norfolk*
★★★**Butterfly Hotel** Beveridge Way, Hardwick Narrows Estate, King's Lynn PE30 4NB 01553-771707 Fax 01553-768027
sB&B £55.50

Oakwood House Hotel *Acclaimed* Tottenhill, King's Lynn PE33 0RH 01553-810256
sB&B £30

KINGSTON UPON THAMES *Surrey*
★★**Hotel Antoinette** 26 Beaufort Road, Kingston upon Thames KT1 2TQ 0181-5461044
Fax 0181-5472595
sB&B £45-£58

KINGTON *Hereford & Worcester*
▲⚏ **Fleece Meadow Caravan Site** Mill Street, Kington HR5 3EQ 01544-430278
Car & Caravan £4.70

KINTBURY *Berkshire*
⚐ ★★★**Elcot Park Resort Hotel** Kintbury RG16 8NJ 01488-58100 Fax 01488-58288
sB&B £82.50

KNUTSFORD *Cheshire*
★★★**Cottons Hotel** Manchester Road, Knutsford WA16 0SU 01565-650333
Fax 01565-755351
sB&B £94

(Inn) The Dog Inn Well Bank Lane, Over Peover, Knutsford WA16 8UP 01625-861421
Fax 01625-861421
sB&B £38

PLACES TO STAY

LAMPLUGH *Cumbria*
▲♥ **Inglenook Caravan Park** Lamplugh CA14 4SH 01946-861240
Car & Caravan £8

LANCASTER *Lancashire*
★★★★*HR* **Lancaster House Hotel** Green Lane, Ellel, Lancaster LA1 4GJ 01524-844322
Fax 01524-844766
Lancaster House successfully combines the finest facilities of a new hotel with the elegance and grace of the finest country house hotel.
& parking, ramp (or level) access, adapted ground floor bedrooms, adapted en-suite bathrooms, adapted bathrooms, sB&B £77-£82

▲♥ **Holgate Caravan Park** Core Road, Silverdale, Lancaster 01524-701508
Fax 01524-701580
Car & Caravan £13.75

LANGPORT *Somerset*
▲♥ **Thorney Lakes Caravan Park** Muchelney., Langport 01458-250811
Car & Caravan £6

LEA MARSTON *Warwickshire*
★★★**Lea Marston Hotel** Haunch Lane, Lea Marston B76 0BY 01675-470468
Fax 01675-470871
sB&B £75

LEAMINGTON SPA *Warwickshire*
★★★**Angel** Regent Street, Leamington Spa CV32 4NZ 01926-881296
sB&B £25.50-£49.50

★**Lansdowne** Clarendon Street, Leamington Spa CV32 4PF 01926-450505
Fax 01926-421313
&ramp (or level) access, wheel chair Lift, sB&B £49.95

LEEDS *West Yorkshire*
★★★★**Hilton National Leeds City** Neville Street, Leeds LS1 4BX 0113-2442000 Fax 0113-2433577
sB&B £99.85

★★★★**Queens Hotel** PO Box 118, City Square, Leeds LS1 1PL 0113-2431323 Fax 0113-2425154
sB&B £74.50

LEEMING BAR *North Yorkshire*
★★★**Motel Leeming** Leeming Bay, Bedale, Northallerton DL8 1DT 01677-422122
Fax 01677-424507
sB&B £35

★★**White Rose** Leeming Bar, Northallerton DL7 9AY 01677-422707 Fax 01677-425123
sB&B £29.50

LEICESTER *Leicestershire*
★★★★**Holiday Inn** St Nicholas Circle, Leicester LE1 5LX 0116-2531161 Fax 0116-2513169
& parking, ramp (or level) access, wheel chair Lift, adapted en-suite bathrooms, sB&B £93.95

★★★**Leicester Moat House** Wigston Road, Oadby, Leicester LE2 5QE 0116-2719441
Fax 0116-2720559
sB&B £60

★★★**Stage Hotel** 299 Leicester Road (A50), Wigston Fields, Leicester LE18 1JW 0116-2886161
Fax 0116-2811874 sB&B £62

★★(Inn) **Red Cow Hotel** Hinckley Road, Leicester Forest East, Leicester LE3 3PG 0116-2387878 Fax 0116-2387878
sB&B £39.50

LENHAM *Kent*
The Harrow Inn *Acclaimed* Warren Street, Lenham ME17 2ED 01622-858727
Fax 01622-850026
sB&B £45

LEWES *East Sussex*
★★**White Hart Hotel** High Street, Lewes BN7 1XE 01273-476694 Fax 01273-476695
sB&B £54

LEYBURN *North Yorkshire*
★(Inn) *C* **Golden Lion Hotel** Market Square, Leyburn DL8 5AS 01969-22161 Fax 01969-23836
&ramp (or level) access, wheel chair Lift, adapted en-suite bathrooms, sB&B £20-£25

LEYLAND *Lancashire*
★★★**Leyland Resort Hotel** (M6 Junction 28), Leyland Way, Leyland PR5 3JX 01772-422922
Fax 01772-622282
sB&B £55

LICHFIELD *Staffordshire*
Coppers End Guest House Walsall Road, Muckley Corner, Lichfield WS14 0BG 01543-372910
sB&B £21

LINCOLN *Lincolnshire*
D'isney Place Hotel *Highly Acclaimed* Eastgate, Lincoln LN2 4AA 01522-528881
Fax 01522-511321
sB&B £49

Hollies Hotel 65 Carholme Road, Lincoln LN1 1RT 01522-522419 Fax 01522-522419
sB&B £25

LISKEARD *Cornwall*
▲♥ **Colliford Tavern** Liskeard PL14 6PZ 01208-821335
Car & Caravan £7

LIVERPOOL Merseyside

★★★★**Atlantic Tower** Chapel Street, Liverpool L3 9RE 0151-2274444 Fax 0151-2363973
sB&B £85

★★★★**Liverpool Moat House** Paradise Street, Liverpool L1 8JD 0151-7090181
Fax 0151-7092706
sB&B £98.25

★★★**Gladstone Hotel** Lord Nelson Street, Liverpool L3 5QB 0151-7097050
Fax 0151-7092193
sB&B £80.95

Hotel Campanile Chaloner Street, Queens Dock, Liverpool L3 4AJ 0151-709-8104
sB&B £29.50

LIZARD Cornwall

★★**Housel Bay Hotel** Housel Cove, Lizard TR12 7PG 01326-290417 Fax 01326-290359
sB&B £28-£32

LONDON (INNER LONDON)

★★★★★**Conrad Hotel** Chelsea Harbour, London SW10 0XG 0171-8233000
Fax 0171-3516525
sB&B £192.25-£280.25

★★★★★**Dorchester Hotel** Park Lane, London W1A 2HJ 0171-629-8888
Fax 0171-409-0114
sB&B £240.63

★★★★★**Four Seasons Hotel** Hamilton Place, Park Lane, London W1A 1AZ 0171-499-0888 Fax 0171-499-5572
sB&B £231.50

★★★★★**Hyde Park Hotel** Knightsbridge, London SW1Y 7LA 0171-2352000
Fax 0171-2354552
sB&B £244

★★★★★**Lanesborough Hotel** 1 Lanesborough Place, London SW1X 7TA 0171-259-5599
sB&B £252.30

★★★★★**London Hilton On Park Lane** Park Lane, London W1Y 4BE 0171-493-8000
Fax 0171-493-4957
sB&B £242.08

★★★★★**London Marriott Hotel** Grosvenor Square, London W1A 4AW 0171-4931232
Fax 0171-491-3201
sB&B £200-£235

★★★★*HC* **Chesterfield Hotel** 35 Charles Street, London W1X 8LX 0171-491-2622
Fax 0171-491-4793
sB&B £96-£126

★★★★**Copthorne Tara** Scarsdale Place, Kensington, London W8 5SR 0171-937 7211
Fax 0171-937 7100
sB&B £105-£110

★★★★**Drury Lane Moat House** 10 Drury Lane, High Holborn, London WC2B 5RE
0171-836-6666 Fax 0171-831-1548
sB&B £120

★★★★*HCR* **Halkin Hotel** Halkin Street, Belgravia, London SW1X 7DJ 0171-333-1000
Fax 0171-333-1100
sB&B £233.25

★★★★**Scandic Crown Hotel** 265 Rotherhithe Street, London SE16 1EJ 0171-231-1001
sB&B £94.50

★★★**Bonnington Hotel** 92 Southampton Row, London WC1B 4BH 0171-242-2828
Fax 0171-831-9170
sB&B £79

Mitre House Hotel *Acclaimed* 178-184 Sussex Gardens, Lancaster Gate, London W2 1TU 0171-7238040 Fax 0171-4020990
sB&B £50-£60

Colonnade Hotel 2 Warrington Crescent, London W9 1ER 0171-2861052 Fax 0171-2861057
sB&B £66

Forte Crest Carburton Street, Regents Park, London W1P 8EE 0171-3882300
Fax 0171-3872806
sB&B £109.95

Forte Crest Coram Street, Bloomsbury, London WC1N 1HT 0171-8371200 Fax 0171-8375374
sB&B £120.95

LONDON (OUTER LONDON)

★★★**Raglan Hall Hotel** Muswell Hill, London N10 3NR 0181-883-9836 Fax 0181-883-5002
sB&B £72

Grove Hill Hotel 38 Grove Hill, South Woodford, London E18 3JG 0181-989-3344
Fax 0181-530-5286
sB&B £23-£35

Sleeping Beauty Motel 543 Lea Bridge Road, Leyton, London E10 7EB 0181-5568080
Fax 0181-5568080
sB&B £35-£40

LONDON AIRPORT-HEATHROW

★★★**Heathrow Park Hotel** Bath Road, West Drayton, London UB7 0EQ 0181-759-2400
Fax 0181-759-5278
sB&B £77.25

PLACES TO STAY

LONG EATON *Derbyshire*
★★★**Novotel Motel** Bostock Lane, Long Eaton, Nottingham NG10 4EP 0115-9720106
Fax 0115-9465900
sB&B £47

Sleep Inn Bostock Lane, Long Eaton NG10 5NL 0115-9460000 Fax 0115-9460726
sB&B £44.45

LONGHORSLEY *Northumberland*
★★★★***HCR* Linden Hall Hotel** Longhorsley NE65 8XF 01670-516611 Fax 01670-788544
sB&B £92.50

LONGTOWN *Cumbria*
★★**(Inn) Graham Arms Hotel** English Street, Longtown CA6 5SE 01228-791213
Fax 01228-791213
sB&B £17-£18.50

LOOE *Cornwall*
Coombe Farm *Highly Acclaimed* Widegates, Looe PL13 1QN 01503-240223
sB&B £18-£24

▲⊕ **Polborder House Caravan & Camping Park** Bucklawren Road, St Martins, Looe PL13 1QR 01503-240265
Car & Caravan £8

LOUTH *Lincolnshire*
★★★★**Kenwick Park Hotel** Kenwick Park, Louth LN11 8NR 01507-608806
sB&B £59.40

★★★**Beaumont Hotel** Victoria Road, Louth LN11 0BX 01507-605005 Fax 01507-607768
&ramp (or level) access, wheel chair Lift, adapted en-suite bathrooms, sB&B £48

LOWESTOFT *Suffolk*
★★★**Victoria Hotel** Kirkley Cliff Road, Lowestoft NR33 0BZ 01505-574433
Fax 01502-501529
sB&B £35-£45.50

LUDLOW *Shropshire*
Cecil Guest House *Acclaimed* Sheet Road, Ludlow SY8 1LR 01584-872442
sB&B £17.50

LUTON *Bedfordshire*
★★★★**Strathmore Thistle Hotel** Arndale Centre, Luton LU1 2TR 01582-34199
Fax 01582-402528
sB&B £83.15

LYME REGIS *Dorset*
★★★***HCR* Alexandra Hotel** Pound Street, Lyme Regis DT7 3HZ 01297-442010 Fax 01297-443229
sB&B £42

★★***H* Bay Hotel** Marine Parade, Lyme Regis DT7 3JQ 01297-442059
sB&B £28-£31

▲⊕ **Shrubbery Caravan Park** Rousdon, Lyme Regis DT7 3XW 01297-442227

LYMINGTON *Hampshire*
▲⊕ **Lytton Lawn Camping & Caravan Park** Lymore, Milford-on-Sea, Lymington 01590-642513
Car & Caravan £18

LYTHAM ST ANNES *Lancashire*
★★★**Chadwick Hotel** South Promenade, Lytham St Annes FY8 1NP 01253-720061
Fax 01253-714455
sB&B £34-£37

MABLETHORPE *Lincolnshire*
▲⊕ **Golden Sands Holiday Park** Quebec Road, Mablethorpe LN12 1QJ 01442-230230
Car & Caravan £4

MACCLESFIELD *Cheshire*
★★**Park Villa Hotel** Park Lane, Macclesfield SK11 8AE 01625-511428 Fax 01625-624637
Victorian Hotel, maintained to the highest standard with three spacious public rooms, both lounges have doors leading to a well maintained garden.
& parking, ramp (or level) access, wheel chair Lift, adapted en-suite bathrooms, sB&B £42

MAIDSTONE *Kent*
★★★★**Tudor Park** Ashford Road, Bearsted, Maidstone ME14 4NQ 01622-734334
Fax 01622-735360
sB&B £75-£85

MALMESBURY *Wiltshire*
⊕ ★★★**Knoll House Hotel** Swindon Road, Malmesbury SN16 9LU 01666-823114
Fax 01666-823897
sB&B £55.50

MALVERN *Hereford & Worcester*
★★★**Abbey Hotel** Abbey Road, Malvern WR14 3ET 01684-892332 Fax 01684-892662
sB&B £65

MALVERN GREAT *Worcestershire*
★★**Great Malvern Hotel** Graham Road, Malvern Great WR14 2HN 01684-563411
Fax 01684-560514
sB&B £45

MANCHESTER *Greater Manchester*
★★★★**Copthorne Hotel** Clippers Quay, Salford Quays Salford, Manchester M5 3DL 0161-8737322 Fax 0161-8737318
&ramp (or level) access, lift, wheel chair Lift, doors, adapted ground floor bedrooms, adapted

en-suite bathrooms, adapted bathrooms, sB&B £112.25

★★★**Novotel** Worsley Brow, Manchester M28 4YA 0161-7993535 Fax 0161-7038207 sB&B £57

Imperial Hotel 157 Hathersage Road, Manchester M13 0HY 0161-225-6500 Fax 0161-225-6500 sB&B £30

MARCH *Cambridgeshire*
★★(Inn) **The Olde Griffin Hotel** High Street, March PE15 9EJ 01354-52517 Fax 01354-50086 sB&B £35

MARGARETTING *Essex*
★★★*H* **Ivy Hill** Writtle Road, Margaretting CM4 0EW 01277-353-040 Fax 01277-355-038 sB&B £65

MARKET HARBOROUGH *Leicestershire*
★★★**Three Swans Hotel** High Street, Market Harborough LE16 7NJ 01858-466644 Fax 01858-433101
Traditional, 16th century, coaching inn situated in a busy town centre. Tastefully refurbished to offer high standards of service and accommodation, and a renowned restaurant.
&ramp (or level) access, adapted en-suite bathrooms, sB&B £67

MARKET RASEN *Lincolnshire*
Walesby Woodlands Walesby, Market Rasen LN8 3UN 01673-843285 sB&B £6

MARKYATE *Hertfordshire*
★★★**Hertfordshire Moat House** London Road, Markyate AL3 8HH 01582-840840 Fax 01582-842282 sB&B £59

MASHAM *North Yorkshire*
▲⚑ **Fearby Caravan Site** Black Swan Hotel, Fearby, Masham HG4 4NF 0176-589477 Car & Caravan £7

MATLOCK *Derbyshire*
▲⚑ **Darwin Forest Country Park** Two Dales, Matlock DE4 5LN 01629-732428 Car & Caravan £10

▲⚑ **Wayside Farm Caravan Site** Matlock Moor, Matlock DE4 5LF 01629-582967 Car & Caravan £5

MEALSGATE *Cumbria*
▲⚑ **Larches Caravan Park** Mealsgate CA5 1LQ 016973-71379 Fax 016973-71782 Car & Caravan £9.25

MELKSHAM *Wiltshire*
★★*HR* **Conigre Farm** Semington Road, Melksham SN12 6BZ 01225-702229 sB&B £40

MERIDEN *West Midlands*
★★★*HC* **Manor Hotel** Main Road, Meriden CV7 7NH 01676-522735 Fax 01676-522186 sB&B £85

MERSEA ISLAND *Essex*
▲⚑ **Waldegraves Farm Holiday Park** West Mersea, Mersea Island CO5 8SE 01206-382898 Fax 01206-385359 Car & Caravan £9

MIDDLE WALLOP *Hampshire*
⚐ ★★★*R* **Fifehead Manor Hotel** Stockbridge, Middle Wallop SO20 8EG 01264-781565 Fax 01264-781400 sB&B £50

MILDENHALL *Suffolk*
★★★**Smoke House Inn Hotel** Beck Row, Mildenhall IP28 8DH 01638-713223 Fax 01638-712202 sB&B £68

MILTON COMMON *Oxfordshire*
★★★**Belfry Hotel** Brimpton Grange, Milton Common OX9 2JW 01844-279381 Fax 01844-279624 sB&B £72.50

MILTON KEYNES *Buckinghamshire*
★★★**Broughton Hotel** Broughton Village, Milton Keynes MK10 9AA 01908-667726 Fax 01908-604844 sB&B £55

★★★**Friendly Hotel** Monks Way, Two Mile Ash, Milton Keynes MK8 8LY 01908-561666 Fax 01908-568303 sB&B £67.90-£81.10

★★★(Inn) **The Wayfarer** Brickhill Street, Milton Keynes sB&B £57.50-£60

Forte Crest 500 Saxon Gate West, Milton Keynes MK9 2HQ 01908-667722 Fax 01908-674714 sB&B £98.50

MODBURY *Devon*
▲⚑ **Pennymoor Camping & Caravan Park** Modbury PL21 0SB 01548-830269 Car & Caravan £11

MOLESWORTH *Cambridgeshire*
(Inn) **Cross Keys** nr Huntingdon, Molesworth PE11 0QF 01832-710283 sB&B £23.25

PLACES TO STAY

MONK FRYSTON *North Yorkshire*
۹ ★★★*HCR* **Monk Fryston Hall Hotel** Selby Road, Monk Fryston LS25 5DU 01977-682369 Fax 01977-683544
sB&B £64-£72

MORECAMBE *Lancashire*
▲♣ **Venture Caravan Park** Langridge Way, Westgate, Morecambe LA4 4TQ 01524-Fax 01524-855884
Car & Caravan £7

MORETON *Merseyside*
★★★*R* **Leasowe Castle Hotel** Leasowe, Moreton L46 3RF 0151-6069191 Fax 0151-6785551
sB&B £39.50

MOTTRAM ST ANDREW *Cheshire*
★★★**Mottram Hall Hotel** Prestbury, Mottram St Andrew SK10 4QT 01625-828135 Fax 01625-828950
sB&B £50-£105

MOULSFORD-ON-THAMES *Oxfordshire*
★★*HCR* **Beetle & Wedge Hotel** Ferry Lane, Moulsford-on-Thames OX10 9JF 01491-651381 Fax 01491-651376
sB&B £80-£100

NAILSWORTH *Gloucestershire*
Apple Orchard House *Highly Acclaimed* Orchard Close, Springhill, Nailsworth GL6 0LX 01453-832503 Fax 01453-836213
sB&B £18-£24

NANTWICH *Cheshire*
۹ ★★★★*CR* **Rookery Hall Hotel** Worleston, Nantwich CW5 6DQ 01270-610016 Fax 01270-626027
sB&B £95

★★**Crown Hotel** High Street, Nantwich CW5 5AS 01270-625283 Fax 01270-628047
sB&B £57.50

★★**(Inn) Malbank Hotel** 14 Beam Street, Nantwich CW5 5LL 01270-626011 Fax 01270-624435
sB&B £28

NEWARK *Nottinghamshire*
(Inn) The Willow Tree Inn Front Street, Barnby in the Willows, Newark NG24 2SA 01636-626913 Fax 01636-626613
sB&B £20

NEWBURY *Berkshire*
★★★★*C* **Donnington Valley Hotel** Old Oxford Road, Newbury RG16 9AG 01635-551199 Fax 01635-551123
sB&B £87.50

★★★★**Regency Park Hotel** Bowling Green Road, Thatcham, Newbury RG13 3RP 01635-873755 Fax 01635-871571
sB&B £83.95

۹ ★★★*CR* **Hollington House Hotel** Woolton Hill, Newbury RG15 9XR 01635-255100 Fax 01635-255075
sB&B £85

★★★*C* **Millwaters Hotel** London Road, Newbury RG13 2BY 01635-528838 Fax 01635-523406
sB&B £59.50

NEWBY BRIDGE *Cumbria*
۹ ★★★*HC* **Lakeside Hotel on Windermere** Newby Bridge LA12 8AT 0153-95-31207 Fax 0153-95-31699
sB&B £55

▲♣ **Newby Bridge Caravan Park** Canny Hill, Newby Bridge LA12 8NF 0153-95-31030
Car & Caravan £7.30

NEWCASTLE UNDER LYME *Staffordshire*
★★**Friendly Stop Inn** Liverpool Road, Newcastle-under-Lyme ST5 9DX 01782-717000 Fax 01782-713669
sB&B £44.50-£63.75

★**The Deansfield** 98 Lancaster Road, Newcastle-under-Lyme ST5 1DS 01782-619040 Fax 01782-627789
sB&B £25

NEWCASTLE UPON TYNE *Tyne & Wear*
★★★★**Copthorne Hotel** The Close, Quayside, Newcastle upon Tyne NE1 3RT 0191-2220333 Fax 0191-2301111
sB&B £108.95

★★★★**Holiday Inn** Great North Road, Seaton Burn, Newcastle upon Tyne NE13 6BP 0191-2365432 Fax 0191-2368091
sB&B £104.95

★★★★*HC* **Swallow Gosforth Park Hotel** High Gosforth Park, Gosforth, Newcastle upon Tyne NE3 5HN 0191-2364111 Fax 0191-2368192
sB&B £98

★★★★**Vermont Hotel** Castle Garth, Newcastle upon Tyne NE1 1RQ 0191-233-1010 Fax 0191-233-1234
sB&B £95

★★★**Imperial Swallow Hotel** Jesmond Road, Jesmond, Newcastle upon Tyne NE2 1PR 0191-2815511 Fax 0191-2818472
sB&B £80

NEWMARKET *Cambridgeshire*
★★★*HC* **Bedford Lodge Hotel** Bury Road,

Newmarket CB8 7BX 01638-663175
Fax 01638-667391
sB&B £65-£68.50

NEWPORT PAGNELL *Buckinghamshire*
★★★**Coach House Hotel** London Road,
Newport Pagnell MK16 0JA 01908-613688
Fax 01908-617335
sB&B £79.50

NEWQUAY *Cornwall*
★★★**Bristol Hotel** Narrowcliff, Newquay TR7
2PQ 01637-875181 Fax 01637-879347
sB&B £45-£50

Priory Lodge Hotel *Acclaimed* 30 Mount Wise,
Newquay TR7 2BH 01637-874111
sB&B £25-£28

▲☗ **Hendra Holiday Park** Newquay TR8 4NY
01637-875778 Fax 01637-879017

▲☗ **Trekenning Manor Tourist Park** Newquay
TR8 4JF 01637-880462 Fax 01637-880462

NEWTON AYCLIFFE *Co Durham*
♋ ★★★★*HCR* **Redworth Hall Hotel** Redworth,
Newton Aycliffe DL5 6NL 01388-772442
Fax 01388-775112
sB&B £95

NIDD *North Yorkshire*
♋ ★★★★*HCR* **Nidd Hall** nr Harrogate, Nidd
HG3 3BN 01423-771598 Fax 01423-770931
&ramp (or level) access, sB&B £95

NORTH PETHERTON *Somerset*
★★★**Walnut Tree Inn** Bridgwater, North
Petherton TA6 6QA 01278-662255
Fax 01278-663946
sB&B £51

NORTH WARNBOROUGH *Hampshire*
(Inn) **Jolly Miller** *Acclaimed* Nr. Odiham, North
Warnborough RG25 1ET 01256-704030
Fax 01256-704030
sB&B £30

NORTHALLERTON *North Yorkshire*
★★★**Sundial Hotel** Darlington Road,
Northallerton DL6 2XF 01609-780525
Fax 01609-780491
sB&B £45

NORTHAMPTON *Northamptonshire*
★★★★**Swallow Hotel** Eagle Drive, Northampton
NN4 0HW 01604-768700 Fax 01604-769011
&adapted ground floor bedrooms, adapted en-suite bathrooms, adapted bathrooms, sB&B £89.50

★★★**Heyford Manor Hotel** Flore, Northampton
NN7 4LP 01327-349022 Fax 01327-349017
sB&B £66.50

▲☗ **Billing Aquadrome** Little Billing,
Northampton NN3 4DA 01604-408181
Fax 01604-784412
Car & Caravan £9.50

NORTHWICH *Cheshire*
♋ ★★★★**Nunsmere Hall Hotel** Tarporley Road,
Oakmere, Northwich CW8 2ES 01606-889100
Fax 01606-889055
sB&B £96.91

Friendly Floatel London Road, Northwich CW9
5HD 01606-44443 Fax 01606-741500
sB&B £58.75

NORWICH *Norfolk*
★★★★**Dunstan Hall** Ipswich Road, Norwich
NR14 5PQ 01508-470444 Fax 01508-471499
sB&B £79.50

★★★**Friendly Hotel** 2 Barnard Road,
Bowthorpe, Norwich NR5 9JB 01603-741161
Fax 01603-741500
sB&B £67.90-£81.10

★★★**Nelson Hotel** Prince Of Wales Road,
Norwich NR1 1DX 01603-760260
Fax 01603-620008
sB&B £73.50

★★★**Norwich Hotel** 121/131 Boundary Road
(A47), Norwich NR3 2BA 01603-787260
Fax 01603-400466
sB&B £59.50

★★★**Norwich Sport Village** Drayton High Road,
Hellesdon, Norwich NR6 5DU 01603-789469
Fax 01603-406845
sB&B £59

NOTTINGHAM *Nottinghamshire*
★★★★**Royal Moat House International**
Wollaton Street, Nottingham NG1 5RH
0115-9414444 Fax 0115-9475667
sB&B £86

★★★**George Hotel** George Street, Nottingham
NG1 3BP 0115-9475641 Fax 0115-9483292
sB&B £64.60-£76.70

★★★**Holiday Inn Garden Court** Castle Marina
Park, Nottingham NG7 1GX 0115-9500600
Fax 0115-9500433
sB&B £67-£72

★★★**Nottingham Moat House** Mansfield Road,
Nottingham NG5 2BT 0115-9602621
Fax 0115-9691506
Modern, three storey, building situated one mile from the city centre.
& parking, ramp (or level) access, lift, wheel chair Lift, adapted ground floor bedrooms, sB&B £63.50

★★★**Stage Hotel** Gregory Boulevard, Nottingham NG7 6LB 0115-9603261 Fax 0115-9691040
sB&B £38

★★★**Strathdon Thistle Hotel** Derby Road, Nottingham NG1 5FT 0115-9418501 Fax 0115-9483725
sB&B £76.95

NUNEATON *Warwickshire*
⚐ **Wolvey Villa Farm Caravan & Camp Site** Wolvey, Nuneaton LE10 3HF 01455-220493
Car & Caravan £4.70

OAKHAM *Leicestershire*
★★*HC* **Boultons Hotel** 4 Catmose Street, Oakham LE15 6HW 01572-722844 Fax 01572-724473
sB&B £35-£50

OKEHAMPTON *Devon*
⚐ **Yertiz Caravan Park** Exeter Road, Okehampton EX20 1QF 01837-52281
Car & Caravan £4

OLDHAM *Gtr Manchester*
★★★*C* **Hotel Smokies Park** Ashton Road, Bardsley, Oldham OL8 3HX 0161-6243405 Fax 0161-6275262
sB&B £52

ORMSKIRK *Lancashire*
⚐ **Abbey Farm Caravan Park** Dark Lane, Ormskirk L40 5TX 01695-572686
Car & Caravan £6

OSWESTRY *Shropshire*
★★★*C* **Wynnstay Hotel** Church Street, Oswestry SY11 2SZ 01691-655261 Fax 01691-670606
sB&B £55

OTLEY *West Yorkshire*
★★★**Chevin Lodge Country Park Hotel** Yorkgate, Otley LS21 3NU 01943-467818 Fax 01943-850335
sB&B £76

OXFORD *Oxfordshire*
✤ ★★**Westwood Country Hotel** Hinksey Hill Top, Oxford OX1 5BG 01865-735408 Fax 01865-736536
sB&B £55

⚐ **Oxford Camping International** 426 Abingdon Road, Oxford OX1 4XN 01865-246551 Fax 01865-240145
Car & Caravan £4.45

OXHILL *Warwickshire*
⚐ **Nolands Farm** *Acclaimed* Oxhill CV35 0RJ 01926-640309 Fax 01926-641662
sB&B £15

PADSTOW *Cornwall*
★★★**Metropole Hotel** Station Road, Padstow PL28 8DB 01841-532486 Fax 01841-532867
sB&B £35-£58

⚐ **Trerethern Touring Park** Padstow PL28 8LE 01841-532061
Car & Caravan £4.80

PADWORTH *Berkshire*
★★★**Padworth Court Hotel** Bath Road, Padworth RG7 5HT 01734-714411 Fax 01734-714442
sB&B £72.50

PAIGNTON *Devon*
★★**Dainton Hotel** 95 Dartmouth Road, Three Beaches Goodrington, Paignton TQ4 6NA 01803-550067 Fax 01803-666339
sB&B £35

★★**Sunhill Hotel** Alta Vista Road, Goodrington Sands, Paignton TQ4 6DA 01803-557532 Fax 01803-663850
sB&B £20.50-£28

★★**Torbay Holiday Motel** Totnes Road, Paignton TQ4 7PP 01803-558226 Fax 01803-663375
sB&B £28.50-£31.70

Sattva Hotel 29 Esplanade, Paignton TQ4 6BL 01803-557820
sB&B £18-£26

⚐ **Byslades Camping & Caravan Park** Totnes Road, Paignton TQ4 7PY 01803-555072
Car & Caravan £9.70

⚐ **Widend Camping Park** Berry Pomeroy Road, Marldon, Paignton TQ3 1RT 01830-550116 Fax 01803-665088

PENRITH *Cumbria*
★★★★**The North Lakes Gateway Hotel** Ullswater Road, Penrith CA11 8QT 01768-68111 Fax 01768-68291
Built in the style of a hunting lodge with open beams and fires, the hotel is an outstanding example of quality, style and comfort. Ideal for touring the Lake District.
♿ parking, lift, wheel chair Lift, doors, adapted ground floor bedrooms, adapted en-suite bathrooms, adapted bathrooms, sB&B £89

⚐ **Gill Head Farm** Troutbeck, Penrith CA11 0ST 01768-79652

⚐ **Park Foot Caravan & Camping Park** Park Foot, Pooley Bridge, Penrith 01768-486309
Car & Caravan £4.50

PENZANCE *Cornwall*
★★★**Queen's Hotel** Promenade, Penzance TR18 4HG 01736-62371 Fax 01736-50033
sB&B £35-£47

PERRANPORTH *Cornwall*
▲⚘ **Perran Sands Holiday Super Centre** Perranporth TR6A OAQ 01442-230300 Fax 01442-230368
Car & Caravan £7.50

PETERBOROUGH *Cambridgeshire*
★★★★*C* **Swallow Hotel** Alwalton Village, Lynch Wood, Peterborough PE2 0GB 01733-371111 Fax 01733-236725
sB&B £88

★★★**Butterfly Hotel** Thorpe Meadows, Off Longthorpe Parkway, Peterborough PE3 6GA 01733-64240 Fax 01733-65538
sB&B £64

PETERCHURCH *Hereford & Worcester*
▲⚘ **Poston Mill Park** Golden Valley, Peterchurch HR2 OSF 01981-550225 Fax 01981-550885
Car & Caravan £6.50

PEVENSEY *East Sussex*
★★**Priory Court Hotel** Pevensey Castle, Pevensey BN24 5LG 01323-763150
sB&B £25-£34

PICKERING *North Yorkshire*
▲⚘ **Wayside Caravan Park** Wrelton, Pickering YO18 8PG 01751-472608
Car & Caravan £6.70

PLYMOUTH *Devon*
★★★★**Plymouth Moat House** Plymouth Hoe, Armada Way, Plymouth PL1 2HJ 01752-662866 Fax 01752-673816
sB&B £66-£86

★★★**New Continental Hotel** Mill Bay Road, Plymouth PL1 3LD 01752-220782 Fax 01752-227013
sB&B £65-£73

PONTEFRACT *West Yorkshire*
★★**Parkside Inn** Park Road, Pontefract WF8 4QD 01977-709911 Fax 01977-701602
sB&B £46-£48

POOLE *Dorset*
★★★★*HR* **Haven Hotel** Sandbanks, Poole BH13 7QL 01202-707333 Fax 01202-708796
♿ parking, ramp (or level) access, lift, wheel chair Lift, good access to restaurant, doors, adapted en-suite bathrooms, adapted bathrooms,
sB&B £75-£85

★★★★**Quay Thistle Hotel** The Quay, Poole BH15 1HD 01202-666800 Fax 01202-684470
sB&B £85

★★★**Harbour Heights Hotel** 73 Haven Road, Poole BH13 7LW 01202-707222 Fax 01202-708594
sB&B £43

★★★*HCR* **Sandbanks Hotel** Banks Road, Poole BH13 7PS 01202-707377 Fax 01202-708885
sB&B £48-£56

Sheldon Lodge Hotel *Acclaimed* 22 Forest Road, Branksome Park, Poole BH13 6DA 01202-761186
sB&B £21-£26

▲⚘ **Beacon Hill Touring Park** Blandford Road North, Near Poole, Poole BH16 6AB 01202-631631
Car & Caravan £6

▲⚘ **Pear Tree Farm Caravan & Camping Park** Organford, Poole BH16 6LA 01202-622434
Car & Caravan £6.25

PORTESHAM *Dorset*
★★**Millmead Country Hotel** Goose Hill, Portesham DT3 4HE 01305-871432
sB&B £33-£37

PORTLAND *Dorset*
Alessandria Hotel 71 Wakeham, Easton, Portland DT5 1HW 01305-822270 Fax 01305-820561
sB&B £30-£35

PORTSCATHO *Cornwall*
⚘ ★★**Roseland House Hotel** Rosevine, Portscatho TR2 5EW 01872-580664 Fax 01872-580801
sB&B £25-£30

PRESTBURY *Cheshire*
★★★**Bridge Hotel** Prestbury SK10 4DQ 01625-829326 Fax 01625-827557
sB&B £76.50

PRESTON *Lancashire*
★★★★**Broughton Park Hotel** Garstang Road, Broughton, Preston PR3 5JB 01772-864087 Fax 01772-861728
sB&B £84

PRINCES RISBOROUGH *Buckinghamshire*
(Inn) George & Dragon 74 High Street, Princes Risborough HP17 0AX 01844-343087 Fax 01844-343087
sB&B £25

QUORN OR QUORNDON *Leicestershire*
★★★★*HCR* **Quorn Country Hotel** Charnwood House, 66 Leicester Road, Quorn LE12 8BB 01509-415050 Fax 01509-415557
sB&B £88.45

RAVENSTONEDALE *Cumbria*
★★**Black Swan Hotel** Kirkby Stephen, Ravenstonedale CA17 4NG 015396-23204 Fax 015396-23604
& parking, ramp (or level) access, adapted en-suite bathrooms, sB&B £42-£46

READING *Berkshire*
★★★★**Holiday Inn** Caversham Bridge, Richfield Avenue, Reading RG1 8BD 01734-391818 Fax 01734-391665
sB&B £115

Abbey House Hotel 118 Connaught Road, Reading 01734-590549 Fax 01734-569299
sB&B £29-£45.50

REDCAR *Cleveland*
★★**Hotel Royal York** 27 Coatham Road, Redcar TS10 1RP 01642-486221 Fax 01642-486221
sB&B £32.50

REDDITCH *Worcestershire*
★★★**Southcrest Hotel** Pool Bank, Mount Pleasant, Redditch B97 4JS 01527-541511 Fax 01527-402600
sB&B £63

Hotel Campanile Far Moor Lane, Winyates Green, Redditch B98 0SD 01527-510710 Fax 01527-517269
sB&B £40

REDMILE *Nottinghamshire*
⚘ **Peacock Farm** *Acclaimed* (In the Vale of Belvoir), Redmile NG13 0GQ 01949-842475 Fax 01949-843127
sB&B £19.50

RICHMOND *North Yorkshire*
Å🚐 **Brompton-On-Swale Caravan Park** Brompton, Richmond DL10 7EZ 01748-824629
&Level caravan access, tent pitches with flat level access, disabled toilet facilities, wheelchair access to reception/office
Car & Caravan £5.20

RICKMANSWORTH *Hertfordshire*
★★★(Inn) **The Long Island Motel** Victoria Close, Rickmansworth WD3 4XQ 01923-775211 Fax 01923-896248
sB&B £63

RINGWOOD *Dorset*
★★*C* **Struan Hotel** Horton Road, Ashley Heath, Ringwood BH24 2EG 01425-473553

Fax 01425-480529
sB&B £35

RIPON *North Yorkshire*
Å🚐 **River Laver Holiday Park** Studley Road, Ripon HG4 2QR 01765-690508
Car & Caravan £8

ROCHESTER *Kent*
★★★★**Bridgewood Manor Hotel** Bridgewood Roundabout, Maidstone Road, Rochester ME5 9AY 01634-201333 Fax 01634-201330
sB&B £85

ROMALDKIRK *Co Durham*
★★*HCR* **Rose & Crown Hotel** Romaldkirk DL12 9EB 01833-50213 Fax 01833-50828
sB&B £53

ROSS-ON-WYE *Hereford & Worcester*
⚘ ★★★*H* **Pengethley Hotel** Ross-on-Wye HR9 6LL 01989-730211 Fax 01989-730238
sB&B £70

★★**Orles Barn Hotel** Wilton, Ross-on-Wye HR9 6AE 01989-562155 Fax 01989-768470
A small family-run hotel situated in a rural setting just off the main road, one mile from the centre of Ross in own secluded grounds of one and a half acres.
&ramp (or level) access, sB&B £30-£45

ROTHERHAM *South Yorkshire*
★★★*C* **Elton Hotel** Main Street, Bramley, Rotherham S66 0SF 01709-545681 Fax 01709-549100
sB&B £52

SALE *Gtr Manchester*
★★★**Amblehurst Hotel** 44 Washway Road, Sale M33 1QZ 0161-9738800 Fax 0161-9051697
sB&B £60

SALFORD *Gtr Manchester*
★★**Hazeldean Hotel** 467 Bury New Road, Kersall, Salford M7 0NX 0161-792-6667 Fax 0161-792-6668
sB&B £34-£44

SALISBURY *Wiltshire*
★★★**Red Lion Hotel** Milford Street, Salisbury SP1 2AN 01722-323334 Fax 01722-325756
sB&B £60

★★★*H* **Rose & Crown Hotel** Harnham Road, Salisbury SP2 8JQ 01722-327908 Fax 01722-339816
sB&B £65

Byways House *Highly Acclaimed* 31 Fowler's Road, Salisbury SP1 2QP 01722-328364 Fax 01722-322146
sB&B £22-£24

▲🚐 **Alderbury Caravan and Camping Park** Old Southampton Road, Whaddon, Salisbury 01722-710125
Car & Caravan £6.50

SAWREY *Cumbria*
★★**The Sawrey Hotel** Far Sawrey, Ambleside, Sawrey LA22 0LQ 015394-43425
sB&B £25.50

SCARBOROUGH *North Yorkshire*
★★**Central Hotel** 1-3 The Crescent, Scarborough YO11 2PW 01723-365766
sB&B £30-£42.50

SCOTCH CORNER *North Yorkshire*
★★★**Friendly Scotch Corner** Scotch Corner DL10 6NR 01748-850900 Fax 01748-825417
sB&B £64.60-£76.70

▲🚐 **Scotch Corner Caravan Park** Richmond, Scotch Corner DL10 6NS 01748-824424
Car & Caravan £3

SELSEY *West Sussex*
▲🚐 **Warner Farm Touring Park** Warner Lane., Selsey 01243-604121
Car & Caravan £9.75

SETTLE *North Yorkshire*
★★★*H* **Falcon Manor** Skipton Road, Settle BD24 9BD 01729-823814 Fax 01729-822087
sB&B £55

SEVENOAKS *Kent*
The Moorings Hotel *Acclaimed* 97 Hitchen Hatch Lane, Sevenoaks TN13 3BE 01732-452589 Fax 01732-456462
sB&B £32

SHAP *Cumbria*
★★★**Shap Wells Hotel** Shap CA10 3QU 01931-716628 Fax 01931-716377
Traditional hotel, situated in a secluded valley high in the Cumbrian Fells. Close to M6, junction 39. The ideal place from which to explore the Lakes, Dales and Border country.
♿ parking, ramp (or level) access, adapted ground floor bedrooms, adapted en-suite bathrooms, sB&B £42

SHEFFIELD *South Yorkshire*
★★★★**Grosvenor House Hotel** Charter Square, Sheffield S1 3EH 0114-2720041
Fax 0114-2757199
sB&B £49.50

★★★*HCR* **Charnwood Hotel** 10 Sharrow Lane, Sheffield S11 8AA 0114-2589411
Fax 0114-2555107
sB&B £75

★★★**Novotel Hotel** Arundel Gate, Sheffield S1 2PR 0114-2781781
sB&B £57

★★★**Rutland Hotel** 452 Glossop Road, Sheffield S10 2PY 0114-2664411 Fax 0114-2670348
sB&B £49

★★★**Sheffield Moat House** Chesterfield Road South, Sheffield S8 8BW 0114-2375376
Fax 0114-2378140
sB&B £79

SHEPTON MALLET *Somerset*
★★★**Charlton House** Charlton Road, Shepton Mallet BA4 4PR 01749-342008 Fax 01749-346362
sB&B £54

SIBSON *Leicestershire*
★★**Millers Hotel** Main Road, Nuneaton, Sibson CV13 6LB 01827-880223 Fax 01827-880223
♿ramp (or level) access, sB&B £41.50

SIDMOUTH *Devon*
★★★★*HCR* **Belmont Hotel** The Esplanade, Sidmouth EX10 8RX 01395-512555
Fax 01395-579101
sB&B £67-£81

★★★★*HC* **Riviera Hotel** The Esplanade, Sidmouth EX10 8AY 01395-515201
Fax 01395-577775
sB&B £56-£66

★★★★*HCR* **Victoria Hotel** Esplanade, Sidmouth EX10 8RY 01395-512651 Fax 01395-579154
sB&B £56-£80

★★★**Fortfield Hotel** Station Road, Sidmouth EX10 8NU 01395-512403 Fax 01395-512403
sB&B £25-£55

Canterbury House Salcombe Road, Sidmouth EX10 8DR 01395-513373
sB&B £16-£18

▲🚐 **Oakdown Touring Park** Weston, Sidmouth EX10 0PH 01395-513731
Car & Caravan £6.50

▲🚐 **Salcombe Regis Camping & Caravan Park** Salcombe Regis, Sidmouth EX10 0JH 01395-514303

SITTINGBOURNE *Kent*
★★★**Coniston Hotel** London Road, Sittingbourne ME10 1NT 01795-472131
Fax 01795-428056
sB&B £49.50

SKEGNESS *Lincolnshire*
▲🚐 **Richmond Holiday Centre** Richmond Drive, Skegness PE25 3TQ 01754-762097

101

Fax 01754-765631
Car & Caravan £7

SKIPTON *North Yorkshire*
★★★**Randell's Hotel** Keighley Road, Snaygill, Skipton BD23 2TA 01756-700100
Fax 01756-700107
sB&B £72

SLOUGH *Berkshire*
★★★★**Copthorne Hotel** Cippenham Lane, Slough SL1 2YE 01753-516222 Fax 01753-516237
sB&B £125

SOLIHULL *West Midlands*
★★★★**Solihull Moat House** Homer Road, Solihull B91 3QD 0121-7114700
Fax 0121-7112696
sB&B £98.20

★★★**Arden Hotel** Coventry Road, Bickenhill, Solihull B92 0EH 01675-443221
Fax 01675-443221
& parking, ramp (or level) access, wheel chair lift, adapted en-suite bathrooms, sB&B £72

★★★*H* **St John's Swallow Hotel** 651 Warwick Road, Solihull B91 1AT 0121-7113000
Fax 0121-7056629
sB&B £85

SOUTH NORMANTON *Derbyshire*
★★★★**Swallow Hotel** Jn 28 (M1), South Normanton DE55 2EH 01773-812000
Fax 01773-580032
sB&B £85

SOUTH SHIELDS *Tyne & Wear*
▲🚐 **Sandhaven Caravan Park** Sea Road, South Shields 0191-4545594
Car & Caravan £8.70

SOUTHAMPTON *Hampshire*
★★★**Novotel** 1 West Quay Road, Southampton SO1 0RA 01703-330550 Fax 01703-222158
sB&B £57

SOUTHEND-ON-SEA *Essex*
★★★**Airport Moat House** Aviation Way, Southend-on-Sea SS2 6UL 01702-546344
Fax 01702-541961
sB&B £55

SOUTHMINSTER *Essex*
▲🚐 **Steeple Bay Caravan Park** Steeple, Southminster CM0 7RS 01442-230300
Fax 01442-230368
Car & Caravan £6.20

SOUTHPORT *Lancashire*
★★★**Scarisbrick Hotel** 239 Lord Street, Southport PR8 1NZ 01704-543000
Fax 01704-533335

& parking, lift, wheel chair lift, good access to restaurant, adapted ground floor bedrooms, adapted en-suite bathrooms, sB&B £50-£65

SPALDING *Lincolnshire*
▲🚐 **Lake Ross Caravan Park** Dozens Bank, West Pinchbeck, Spalding PE11 3NA 01775-761690
Car & Caravan £7

ST AGNES *Cornwall*
& ★★*H* **Rose-in-Vale Country House Hotel** Mithian, St Agnes TR5 0QD 01872-552202
Fax 01872-552700

sB&B £35.50

ST AUSTELL *Cornwall*
★★★★*H* **Carlyon Bay Hotel** Sea Road, Carlyon Bay, St Austell PL25 3RD 01726-812304
Fax 01726-814938
sB&B £62-£75

Alexandra Hotel 52-54 Alexandra Road, St Austell PL25 4QN 01726-74242
sB&B £26

Selwood House 60 Alexandra Road, St Austell PL25 4QN 01726-65707 Fax 01726-68951
sB&B £34

▲🚐 **Carlyon Bay Camping Park** Cypress Avenue, Bethesda, St Austell PL25 3RE 01726-812735
Car & Caravan £12

▲🚐 **Sea View International** Boswinger, Gorran, St Austell PL26 6LL 01726-843425
Car & Caravan £13.70

ST IVES, *Cambridgeshire*
★★★**Dolphin Hotel** Bridge Foot, London Road, St Ives PE17 4EP 01480-466966 Fax 01480-495597
sB&B £55

★★★*H* **Olivers Lodge Hotel** Needingworth Road, St Ives PE17 4JP 01480-463252
Fax 01480-461150
sB&B £48

ST MELLION Cornwall
★★★**St Mellion Golf & Country Club** Saltash, St Mellion PL12 6SD 01579-50101 Fax 01579-50116 sB&B £40-£57

STAFFORD Staffordshire
★★★**Garth Hotel** Moss Pit, Stafford ST17 9JR 01785-56124 Fax 01785-55152 sB&B £61

★★★*H* **Tillington Hall Hotel** Eccleshall Road, Stafford ST16 1JJ 01785-53531 Fax 01785-59223 sB&B £80

STALHAM Norfolk
★★*H* **Kingfisher Hotel** High Street, Stalham NR12 9AN 01692-581974 Fax 01692-582544 sB&B £36

STAMFORD Lincolnshire
★★★**Garden House Hotel** St Martin's, Stamford PE9 2LP 01780-63359 Fax 01780-63339 sB&B £49.75-£62.25

★★★**Lady Anne's Hotel** 37-38 High Street, St Martin, Stamford PE9 1FG 01780-53175 Fax 01780-65422
& parking, adapted ground floor bedrooms, adapted en-suite bathrooms, sB&B £35-£49.50

STANDISH Gtr Manchester
★★★**Almond Brook Moat House** Almond Brook Road, Standish WN6 0SR 01257-425588 Fax 01257-427327 sB&B £52.50-£72.50

STEEPLE ASTON Oxfordshire
Westfield Farm Motel Acclaimed The Fenway, Steeple Aston OX5 3SS 01869-40591 sB&B £32-£36

STEVENAGE Hertfordshire
★★★**Novotel** Knebworth Park, Stevenage SG1 2AX 01438-742299 Fax 01438-723872 sB&B £57

STEYNING West Sussex
★★★*H* **Old Tollgate Hotel** The Street, Bramber, Steyning BN44 3WE 01903-879494 Fax 01903-813399
An old Toll house, with a modern extension, housing luxuriously appointed bedrooms. Masses of olde worlde charm and a popular restaurant.
& ramp (or level) access, wheel chair lift, adapted en-suite bathrooms, sB&B £57.50

STOCKTON-ON-TEES Cleveland
★★★★**Swallow Hotel** 10 John Walker Square, Stockton-on-Tees TS18 1AQ 01642-679721 Fax 01642-601714 sB&B £85

STOKE CANON Devon
★★★*HC* **Barton Cross Hotel** Huxham, Stoke Canon EX5 4EJ 01392-841245 Fax 01392-841942 sB&B £63.50

STOKE-ON-TRENT Staffordshire
★★★★**Stoke on Trent Moat House** Etruria Hall, Festival Park, Etruria, Stoke-on-Trent ST1 5BQ 01782-219000 Fax 01782-284500 sB&B £85

Hanchurch Manor Country House *Highly Acclaimed* Hanchurch, Stoke-on-Trent ST4 8SD 01782-643030 Fax 01782-643035 sB&B £65

STONE Staffordshire
★★★*H* **Stone House Hotel** Stone ST15 0BQ 01785-815531 Fax 01785-814764 sB&B £65

STRATFORD-UPON-AVON Warwickshire
★★★★**Moat House International Hotel** Bridgefoot, Stratford-upon-Avon CV37 6YR 01789-414411 Fax 01789-298589 sB&B £50-£102.75

& ★★★★*HCR* **Welcombe Hotel** Warwick Road, Stratford-upon-Avon CV37 0NR 01789-295252 Fax 01789-414666 sB&B £95-£115

★★★*HC* **Windmill Park Hotel** Warwick Road, Stratford-upon-Avon CV37 0PY 01789-731173 Fax 01789-731131 sB&B £82.50

Sequoia House Hotel *Highly Acclaimed* 51-53, Shipston Road, Stratford-upon-Avon CV57 7LN 01789-268852 Fax 01789-414559 sB&B £29-£49

Nando's *Acclaimed* 18 & 19 Evesham Place, Stratford-upon-Avon CV37 6HT 01789-204907 Fax 01789-204907 sB&B £17-£20

STREATLEY-ON-THAMES Berkshire
★★★★**The Swan Diplomat** Streatley-on-Thames RG8 9HR 01491-873737 Fax 01491-872554 sB&B £96-£105.50

STREET Somerset
★★★**Wessex Hotel** High Street, Street BA16 0EA 01458-43383 Fax 01458-46589 sB&B £40-£50

STROUD Gloucestershire
★★*C* **Bell Hotel** Wallbridge, Stroud GL5 3JA 01453-763556 sB&B £28

STUDLAND *Dorset*
★★★**Knoll House Hotel** Ferry Road, Studland BH19 3AH 01929-44251 Fax 01929-44423
sB&B £45-£72

SUNDERLAND *Tyne & Wear*
★★★★**Swallow Hotel** Queens Parade, Sunderland SR6 8DB 0191-5292041
Fax 0191-5294545
sB&B £90

SUTTON *Gtr London*
The Dene Hotel 39 Cheam Road, Sutton SM1 2AT 0181-6423170 Fax 0181-6423170
sB&B £21-£40

SUTTON COLDFIELD *Warwickshire*
🍴 ★★★★**New Hall Hotel** Walmley Road, Sutton Coldfield B76 8QX 0121-3782442
Fax 0121-3784637
sB&B £100

★★★★*HC* **The Belfry Hotel** Lichfield Road, Wishaw, Sutton Coldfield B76 6BR 01675-470301
Fax 01675-470178
sB&B £115

★★★**Marston Farm Hotel** Bodymoor Heath, Sutton Coldfield B76 9JD 01827-872133
Fax 01827-875043
sB&B £49.50-£70

SWANAGE *Dorset*
★★★*CR* **Pines Hotel** Burlington Road, Swanage BH19 1LT 01929-425211 Fax 01929-422075
♿ parking, ramp (or level) access, lift, wheel chair lift, good access to restaurant, doors, adapted en-suite bathrooms, sB&B £32-£40

SWINDON *Wiltshire*
★★★★*H* **Blunsdon House Hotel** Blunsdon, Swindon SN2 4AD 01793-721701
Fax 01793-721056
sB&B £75-£77

★★★★*C* **De Vere Hotel** Shaw Ridge Leisure Park, Whitehall Way, Swindon SN5 7DW 01793-878785 Fax 01793-877822
sB&B £100

★★★*HCR* **Pear Tree at Purton** Church End, Purton, Swindon SN5 9ED 01793-772100
Fax 01793-772369
sB&B £92

★★★**Stanton House Hotel** The Avenue, Stanton Fitzwarren, Swindon SW6 7SD 01793-861777
Fax 01793-861857
sB&B £65-£90

TAMWORTH *Staffordshire*
⛺ **Drayton Manor Park** Fazeley, Tamworth B78 3TW 01827-287979 Fax 01827-288916
Car & Caravan £10

TATTERSHALL *Lincolnshire*
⛺ **Tattershall Park Country Club** Sleaford Road, Tattershall LN4 4LR 01526-343193
Fax 01526-343308
Car & Caravan £7

TAUNTON *Somerset*
★★★**Castle Hotel** Castle Green, Taunton TA1 1NF 01823-272671
Fax 01823-336066
sB&B £65

★★★**Rumwell Manor Hotel** Wellington Road, Rumwell, Taunton TA4 1EL 01823-461902
Fax 01823-254861
sB&B £49

Meryan House Hotel *Highly Acclaimed* Bishops Hull, Taunton TA1 5EG 01823-337445
Fax 01823-322355
sB&B £36

TAVISTOCK *Devon*
⛺ **Higher Longford Farm** Moorshop, Tavistock PL19 9LQ 01822-613360
Car & Caravan £5.75

TELFORD *Shropshire*
★★★**Telford Hotel, Golf & Country Club** Great Hay, Sutton Hill, Telford TF7 4DT 01952-585642
Fax 01952-586602
Telford Hotel, golf and country club combined, stands high above the splendour of the Ironbridge Gorge and its historic bridge. Situated four miles south of junction 4, of the M54.
♿ parking, ramp (or level) access, adapted en-suite bathrooms, sB&B £91

TEMPLE SOWERBY *Cumbria*
★★★*H* **Temple Sowerby House Hotel** Temple Sowerby CA10 1RZ 0176-83-61578
Fax 0176-83-61958
sB&B £50

TETBURY *Gloucestershire*
🍴 ★★★*HCR* **Calcot Manor** Tetbury GL8 8YG 01666-890391 Fax 01666-890394
sB&B £75

TEWKESBURY *Gloucestershire*
★★★**Royal Hop Pole Hotel** Church Street, Tewkesbury GL20 5RT 01684-293236
Fax 01684-293380
sB&B £83.50

THORNE *South Yorkshire*
★★**Belmont Hotel** Horse Fair Green, Thorne DN4 5EE 01405-812320 Fax 01405-740508
sB&B £46.95-£53.50

THORNLEY *Co Durham*
★★**Crossways Hotel** Dunelm Road, Thornley D6 3HT 01429-821248 Fax 01429-820034
sB&B £25-£37

THORNTON HOUGH *Merseyside*
★★★**Thornton Hall Hotel** Neston Road, Thornton Hough L63 1JF 0151-3363938 Fax 0151-3367864
sB&B £63

TINTAGEL *Cornwall*
★★**Bossiney House Hotel** Bossiney, Tintagel PL34 0AX 01840-770240 Fax 01840-770501
sB&B £24-£29

TIVERTON *Devon*
★★★**Tiverton Hotel** Blundells Road, Tiverton EX16 4DB 01884-256120 Fax 01884-258101
sB&B £29-£38

★★**Hartnoll Country Hotel** Bolham Road, Bolham, Tiverton EX16 7RA 01884-252777 Fax 01884-259195
sB&B £35

▲⚏ **Minnows Camping & Caravan Park** Sampford Peverell, Tiverton EX16 9LD 01884-821770
Car & Caravan £3

TONBRIDGE *Kent*
Chimneys Motor Inn Pembury Road, Tonbridge TN11 0NA 01732-773111
sB&B £36

TORPOINT *Cornwall*
Whitsand Bay Hotel *Acclaimed* Portwinkle, Crafthole, Torpoint PL11 3BU 01503-30276 Fax 01503-30329
sB&B £20-£25

TORQUAY *Devon*
★★★★★**Imperial** Parkhill Road, Torquay TQ1 2DG 01803-294301 Fax 01803-298293
Built in 1866 and renovated on several occasions, hotel is not adequately equipped to deal for the needs of disabled guests.
♿ parking, ramp (or level) access, wheel chair lift, good access to restaurant, adapted ground floor bedrooms, sB&B £90-£140

★★★**HC Corbyn Head Hotel** Sea Front, Torbay Road, Torquay TQ2 6RH 01803-213611 Fax 01803-296152
sB&B £40-£70

★★★**Livermead House Hotel** Sea Front, Torquay TQ2 6QJ 01803-294361 Fax 01803-200758
sB&B £35-£55

★**HR Fairmount House** Herbert Road, Chelston, Torquay TQ2 6RW 01803-605446 Fax 01803-605446
sB&B £26.50-£29

★**Shelley Court Hotel** Croft Road, Torquay TQ2 5UD 01803-295642 Fax 01803-215793
sB&B £18-£24

Craig Court Hotel *Acclaimed* 10 Ash Hill Road, Torquay TQ1 3HZ 01803-294400
sB&B £16.50-£22.50

Lindens Hotel *Acclaimed* 31 Bampfylde Road, Torquay TQ2 5AY 01803-212281
sB&B £15-£18

Lindum Hotel Abbey Road, Torquay TQ2 5ND 01803-292795
sB&B £12-£20

TRICKETTS CROSS *Dorset*
★★**Coach House** Ferndown, Tricketts Cross BH22 9NW 01202-861222 Fax 01202-894130
sB&B £45

TRING *Hertfordshire*
★★★★**HC Pendley Manor Hotel** Cow Lane, Tring HP23 5QY 01442-891891 Fax 01442-890687
sB&B £85

TROWBRIDGE *Wiltshire*
The Old Manor Hotel Trowle, Trowbridge BA14 9BL 01225-777393 Fax 01225-765443
sB&B £45-£50

TUNBRIDGE WELLS *Kent*
★★★★**Spa Hotel** Langton Road, Mount Ephraim, Tunbridge Wells TN4 8XJ 01892-520331 Fax 01892-510575
sB&B £69

★★★**Kingswood Hotel** Pembury Road, Tunbridge Wells TN2 3QS 01892-535736 Fax 01892-513321
sB&B £50-£52

TURVEY *Bedfordshire*
★★**Laws Hotel** High Street, Turvey MK43 8DB 01234-881213 Fax 01234-888864
sB&B £39.50-£42

TUTBURY *Staffordshire*
★★★**R Ye Olde Dog & Partridge** High Street, Tutbury DE13 9LS 01283-813030 Fax 01283-813178
sB&B £52.50

TYNEMOUTH *Tyne & Wear*
Hope House *Highly Acclaimed* 47 Percy Gardens, Tynemouth NE30 4HH 0191-2571989 Fax 0191-2571989
sB&B £35

ULLSWATER *Cumbria*

▲⚘ **Cove Caravan & Camping Park**
Watermillock, Ullswater CA11 0LS 0176-84-86549
Car & Caravan £4.40

▲⚘ **Hillcroft Park** Pooley Bridge, Ullswater
CA10 2LT 0176-84-86363
Car & Caravan £2

▲⚘ **Park Foot Caravan & Camping Park**
Howtown Road, Pooley Bridge, Ullswater CA10
2NA 017684-86309 Fax 017684-86309
Car & Caravan £4.50

UPPINGHAM *Leicestershire*

★★★(Inn) **Marquess of Exeter Hotel** 52 Main
Street, Lyddington, Uppingham LE15 9LT
01572-822477 Fax 01572-821343
sB&B £45

UTTOXETER *Staffordshire*

▲⚘ **Uttoxeter Racecourse Caravan Club Site**
Wood Lane, Uttoxeter ST14 8BD 01889-564172

WADEBRIDGE *Cornwall*

▲⚘ **St Minver Holiday Park** St Minver,
Wadebridge PL27 6RR 01442-230300
Fax 01442-230368
Car & Caravan £7

WAKEFIELD *West Yorkshire*

★★★**St Pierre Hotel** Barnsley Road,
Newmillerdam, Wakefield WF2 6QG
01924-255596 Fax 01924-252746
sB&B £53.20

★★★**Swallow Hotel** Queens Street, Wakefield
WF1 1JU 01924-37211 Fax 01924-383648
sB&B £78

WALSALL *West Midlands*

★★★**Friendly Hotel** Wolverhampton Road West,
Bentley, Walsall WS2 0BS 01922-724444
Fax 01922-723148
sB&B £67.90-£81.10

★★**Abberley Hotel** 29 Bescot Road, Walsall WS2
9AD 01922-27413 Fax 01922-720933
Tastefully refurbished Victorian building maintaining character and style of its period.
&ramp (or level) access, adapted en-suite
bathrooms, sB&B £36

WALTHAM ABBEY *Essex*

★★★★**Swallow Hotel** Old Shire Lane, Waltham
Abbey EN9 3LX 01992-717170 Fax 01992-711841
sB&B £96

WANSFORD *Cambridgeshire*

★★★*CR* **Haycock Hotel** Great North Road,
Wansford PE8 6JA 01780-782223
Fax 01780-783031
sB&B £68

WARE *Hertfordshire*

⚘ ★★★★**Hanbury Manor** Ware SG12
0SD 01920-487722 Fax 01920-487692
sB&B £105-£145

WAREHAM *Dorset*

★★★**Springfield Country Hotel** Grange Road,
Stoborough, Wareham BH20 5AL 01929-552177
Fax 01929-551862
sB&B £58-£65

WARKWORTH *Northumberland*

★★**Warkworth House Hotel** 16 Bridge Street,
Warkworth NE65 0XB 01665-711276
Fax 01665-713323
sB&B £49

WARRINGTON *Cheshire*

★★★*HC* **De Vere Lord Warrington** Daresbury,
Warrington WA4 4BB 01925-267331
Fax 01925-265615
sB&B £95

★★★**Fir Grove Hotel** Knutsford Old Road,
Grappenhall, Warrington WA4 2LD
01925-267471 Fax 01925-601092
sB&B £59

WARWICK *Warwickshire*

★★★★**Hilton National** Stratford Road, Warwick
CV34 6RE 01926-499555 Fax 01926-410020
sB&B £89

WASHINGTON *Tyne & Wear*

★★★**Washington Moat House** Stone Cellar
Road, Usworth Village, Washington NE37 1PH
0191-4172626 Fax 0191-4151166
sB&B £67

WATCHET *Somerset*

▲⚘ **Doniford Holiday Park** Watchet TA23 0TJ
01984-32423
Car & Caravan £6.50

WATERMILLOCK *Cumbria*

⚘ ★★★**Leeming House** Ullswater, Watermillock
CA11 0JJ 0176-84-86622 Fax 0176-84-86443
sB&B £84.50-£105.50

WATFORD *Hertfordshire*

★★**White House Hotel** Upton Road, Watford
WD1 2EL 01923-237316 Fax 01923-233109
sB&B £60

WEEDON BEC *Northamptonshire*

★★(Inn) *C* **Globe Hotel** High Street, Weedon
Bec NN7 4QD 01327-40336 Fax 01327-349058
sB&B £42

WELLS *Somerset*

▲⚘ **Homestead Park** Wookey Hole, Wells BA5

1BW 01749-673022
Car & Caravan £7.80

WEMBLEY *Gtr London*

Adelphi Hotel *Acclaimed* 4 Forty Lane, Wembley HA9 9EB 0181-904-5629 Fax 0181-908-5314
Clean and tidy, pleasant to the eye - noticeable from quite some distance, enjoys prominent position in Wembley. Being near major landmarks spots, yet still enjoying set back and secluded position.
sB&B £28-£35

Arena Hotel *Acclaimed* 6 Forty Lane, Wembley HA9 9EB 0181-9040019 Fax 0181-9082007
sB&B £25-£28

Wembley Park Hotel *Acclaimed* 8 Forty Lane, Wembley HA9 9EB 0181-9046329
Fax 0181-3850472
sB&B £28-£32

Brookside Hotel 32 Brook Avenue, Wembley HA9 8PH 0181-9040252 Fax 0181-3850800
sB&B £25

WENTBRIDGE *West Yorkshire*

★★★**Wentbridge House Hotel** Nr Pontefract, Wentbridge WF8 3JJ 01977-620444
Fax 01977-620148
sB&B £45-£68.25

WESTON-SUPER-MARE *Avon*

★★★**HR Commodore Hotel** Beach Road, Sand Bay Kewstoke, Weston-super-Mare B22 9UZ 01934-415778 Fax 01934-636483
sB&B £49

★★**Arosfa Hotel** Lower Church Road, Weston-super-Mare BS23 2AG 01934-419523
Fax 01934-636084
sB&B £35-£38.50

★★**Beachlands Hotel** 17 Uphill Road North, Weston-super-Mare BS23 4NG 01934-621401
Fax 01934-621966
Situated overlooking championship 18 hole golf course, 200 yards level walk from seafront with easy access to town centre.
♿ramp (or level) access, adapted en-suite bathrooms, sB&B £28.50-£31

L'Arrivee Guest House *Acclaimed* 75 Locking Road, Weston-super-Mare BS23 3DW
01934-625328
♿ parking, ramp (or level) access, doors, adapted ground floor bedrooms, adapted en-suite bathrooms,
sB&B £15-£16

Saxonia Guest House *Acclaimed* 95 Locking Road, Weston-super-Mare BS23 3EW
01934-633856
sB&B £18-£23

▲⊞ **West End Farm Caravan Park** Locking, Weston-super-Mare BS24 8RH 01934-822529
Car & Caravan £8.80

WESTONBIRT *Gloucestershire*

★★★**Hare & Hounds Hotel** Tetbury, Westonbirt GL8 8QL 01666-880233 Fax 01666-880241
sB&B £58-£68

WETHERAL *Cumbria*

★★★**Crown Hotel** Wetheral, Carlisle CA4 8ES 01228-561888 Fax 01228-561637
This elegant Grade II listed Georgian building stands next to the banks of the River Eden, east of Carlisle. There's a traditional bar and a conservatory restaurant overlooking the floodlit gardens.
♿ramp (or level) access, adapted en-suite bathrooms, sB&B £88

WEYMOUTH *Dorset*

★★**Central Hotel** 15 Maiden Street, Weymouth DT4 8BB 01305-760700 Fax 01305-760300
sB&B £25-£28

★★**Crown Hotel** 51/52 St Thomas Street, Weymouth DT4 8EQ 01305-760800
Fax 01305-760300
sB&B £28.50-£32

★★**Moonfleet Manor Hotel** Moonfleet, Weymouth DT3 4ED 01305-786948
Fax 01305-774395
sB&B £30-£35

★**Alexandra** 27/28 The Esplanade, Weymouth DT4 8DN 01305-785767
sB&B £19.50

Sou West Lodge Hotel Rodwell Road, Weymouth DT4 8QT 01305-783749
sB&B £20.25-£21.90

▲⊞ **Bagwell Farm Touring Park** Chickerell, Weymouth DT3 4EA 01305-782575
Car & Caravan £8.75

▲⊞ **East Fleet Farm Touring Park** Chickerell, Weymouth DT3 4DW 01305-785768
Car & Caravan £7.50

▲⊞ **Weymouth Bay Holiday Park** Preston, Weymouth DT3 6BQ 01422-230300
Fax 01442-230368
Car & Caravan £7.50

WHATSTANDWELL *Derbyshire*

▲⊞ **Merebrook Caravan Park** Derby Road, Whatstandwell DE4 5HH 01773-852154
Car & Caravan £8

WHATTON *Nottinghamshire*

★★**(Inn) Haven Hotel** Grantham Road (A52), Whatton NG13 9EU 01949-50800

Fax 01949-51454
sB&B £30

WHITBY *North Yorkshire*
Dunsley Hall Hotel *Highly Acclaimed* Dunsley, Whitby YO21 3TL 01947-893437
Fax 01947-893505
sB&B £36.50-£45

🏕️ **Northcliffe Holiday Park** High Hawsker, Whitby YO22 4LL 01947-880477
Car & Caravan £13

🏕️ **York House** High Hawsker, Whitby YO22 4LW 01947-880354
Car & Caravan £6.50

WHITCHURCH *Shropshire*
★★**Redbrook Hunting Lodge Hotel** Wrexham Road, Whitchurch SY13 3ET 01948-73204
Fax 01948-73533
sB&B £35-£40

WHITLEY BAY *Tyne & Wear*
★★**Seacrest** 14-18 North Parade, Whitley Bay NE26 1PA 0191-253-0140 Fax 0191-253-0140
sB&B £28

York House Hotel *Acclaimed* 30 Park Parade, Whitley Bay NE26 1DX 0191-252-8313
Fax 0191-252-8313
sB&B £26.50

WICKFORD *Essex*
★★★**Chichester Hotel** Old London Road, Wickford SS11 8UE 01268-560555
Fax 01268-560580
sB&B £58.45

WIGAN *Gtr Manchester*
★★★**Wigan Oak Hotel** River Way, Wigan WN1 3SS 01942-826888 Fax 01942-825800
sB&B £61.75

WIGGLESWORTH *North Yorkshire*
★★★**(Inn) Plough Inn** Wigglesworth BD23 4RJ 01729-840243
Delightful 18th-century inn with beams and open fires, set in beautiful countryside with panoramic views of the peaks.
♿ parking, ramp (or level) access, sB&B £28.85-£32.45

WILLITON *Somerset*
★★**HCR White House Hotel** Long Street, Williton TA4 4QW 01984-632306
sB&B £39-£45

WIMBORNE MINSTER *Dorset*
🏕️ **Springfield Touring Park** Candys Lane, Corfe Mullen, Wimborne Minster BH21 3EF 01202-881719
Car & Caravan £7

WINCANTON *Somerset*
🏕️ **Sunnyhill Farm** Wincanton 01963-33281
Car & Caravan £5.50

WINCHESTER *Hampshire*
❦ ★★★★**HCR Lainston House Hotel** Sparsholt, Winchester SO21 2LT 01962-863588
Fax 01962-776672
sB&B £105

★★★**Winchester Moat House** Worthy Lane, Winchester SO23 7AB 01962-868102
Fax 01962-840862
sB&B £57.75-£78.25

Shawlands *Acclaimed* 46 Kilham Lane, Winchester SO22 5QD 01962-861166
Fax 01962-861166
sB&B £20

WINDERMERE *Cumbria*
★★★**Belsfield Hotel** Kendal Road, Windermere LA23 3EL 015394-42448 Fax 015394-46397
sB&B £83-£93

★★★**H Burn How Garden House Hotel** Belsfield Road, Bowness, Windermere LA23 3HH 0153-94-46226 Fax 0153-94-47000
sB&B £46-£52

★★★**Burnside Hotel** Kendal Road, Bowness, Windermere LA23 3EP 0153-94-42211
Fax 0153-94-43824
sB&B £47.50

❦ ★★★**Langdale Chase Hotel** Windermere LA23 1LW 0153-94-32201 Fax 0153-94-32604
sB&B £42

❦ ★★★**HCR Linthwaite House** Crook Road, Windermere LA23 3JA 0153-94-88600
Fax 0153-94-88601
sB&B £69

★★★**Low Wood Hotel** Windermere LA23 1LP 0153-94-33338 Fax 0153-94-34072
sB&B £47-£49.50

❦ ★★**HCR Lindeth Fell Hotel** Windermere LA23 3JP 015394-43286
sB&B £41

Broad Oak Country House *Highly Acclaimed* Bridge Lane, Troutbeck, Windermere LA23 1LA 015394-45566 Fax 015394-88766
sB&B £35

Hawksmoor *Highly Acclaimed* Lake Road, Windermere LA23 2EQ 015394-42110
sB&B £20-£30

WINDSOR *Berkshire*
Netherton Hotel *Highly Acclaimed* 96-98 St Leonards Road, Windsor SL4 3NU 01753-855508

A three storey, red-brick hotel, ten minutes from the centre of Windsor.
& ramp (or level) access, adapted en-suite bathrooms, sB&B £30

WINSLOW *Buckinghamshire*
★★(Inn) **The Bell Hotel** Market Square, Winslow MK18 3AB 01296-714091 Fax 01296-714805
sB&B £34.50

WIX *Essex*
🛥 **New Farm** *Acclaimed* Spinnell's Lane, Manningtree, Wix CO11 2UJ 01255-870365 Fax 01255-870837
sB&B £19.50

WOKINGHAM *Berkshire*
Å♣ **California Chalet & Touring Park** Nine Mile Ride, Finchampstead, Wokingham RG11 3NY 01734-733928
Car & Caravan £8.25

WOLVERHAMPTON *West Midlands*
★★★**Novotel** Union Street, Wolverhampton WV1 3JN 01902-871100 Fax 01902-870054
sB£B £49.50

WOMENSWOLD *Kent*
Woodpeckers Country Hotel Womenswold CT4 6HB 01227-831319 Fax 01227-831319
sB&B £26

WOODBRIDGE *Suffolk*
★★★**Ufford Park Hotel** Yarmouth Road, Ufford, Woodbridge IP12 1QW 01394-383555 Fax 01394-383582
sB&B £60-£70

WOODHALL SPA *Lincolnshire*
★★★**Golf Hotel** The Broadway, Woodhall Spa LN10 6SG 01526-353535 Fax 01526-353096
A Tudor style golf hotel furnished to high standards of modern comfort, whilst maintaining the building's traditional character. Set in acres of gardens and woodland.
& parking, ramp (or level) access, doors, adapted ground floor bedrooms, adapted en-suite bathrooms, sB&B £35-£40

Å♣ **Bainland Country Park** Horncastle Road, Woodhall Spa LN10 6UX 01526-352903 Fax 01526-353730

WOOLACOMBE *Devon*
★★*H* **Headlands Hotel** Beach Road, Woolacombe EX34 7BT 01271-870320
sB&B £20-£28

WOOLER *Northumberland*
Å♣ **Riverside Caravan Park** Wooler NE71 6EE 01442-230300 Fax 01442-230368
Car & Caravan £6.50

WORCESTER *Hereford & Worcester*
★★★*HC* **Fownes Resort Hotel** 5 Clare Street, City Walls Road, Worcester WR1 2AP 01905-613151 Fax 01905-23742
sB&B £82.50

★★★**Star Hotel** Foregate Street, Worcester WR1 1EA 01905-24308 Fax 01905-23440
sB&B £49.50

Å♣ **Mill House Caravan & Camping Site** Hawford, Worcester WR3 7SE 01905-451283
Car & Caravan £4.50

WORKINGTON *Cumbria*
Morven Hotel *Highly Acclaimed* Siddick Road, Workington CA14 1LE 01900-602118
sB&B £20-£29

WORKSOP *Nottinghamshire*
★★★**Charnwood Hotel** Sheffield Road, Blyth, Worksop S81 8HF 01909-591610 Fax 01909-591429
sB&B £39.50

★★★**Clumber Park Hotel** Clumber Park, Worksop S80 3PA 01623-835333 Fax 01623-835525
sB&B £66.50

Å♣ **Riverside Caravan Park** Central Avenue, Worksop S80 1ER 01909-474118
Car & Caravan £4.25

WORTHING *West Sussex*
★★★**Chatsworth Hotel** Steyne, Worthing BN11 3DU 01903-236103 Fax 01903-823726
sB&B £49.90-£59.90

★★★**Windsor House Hotel** 14/20 Windsor Road, Worthing BN11 2LX 01903-239655 Fax 01903-210763
sB&B £35

★★**Ardington Hotel** Steyne Gardens, Worthing BN11 3DZ 01903-230451 Fax 01903-230451
sB&B £50

Delmar Hotel *Acclaimed* 1-2 New Parade, Worthing BN11 2BQ 01903-211834 Fax 01903-850249
sB&B £26

YEOVIL *Somerset*
★**Preston Hotel** 64 Preston Road, Yeovil BA20 2DL 01935-74400 Fax 01935-410142
sB&B £30-£39

YORK *North Yorkshire*

★★★★*HC* **Royal York Hotel** Station Road, York YO2 2AA 01904-653681 Fax 01904-623503
sB&B £87.50

★★★★**Swallow Hotel** Tadcaster Road, Dringhouses, York YO2 2QQ 01904-701000 Fax 01904-702308
A disabled-guest-friendly hotel. Located overlooking York Racecourse. E.T.B. Category one establishment. All public areas have level access.
& parking, ramp (or level) access, lift, wheel chair Lift, good access to restaurant, doors, adapted ground floor bedrooms, adapted en-suite bathrooms, sB&B £88

★★★★**Viking Hotel** North Street, York YO1 1JF 01904-659822 Fax 01904-641793
sB&B £98.50-£115

★★★*HC* **Fairfield Manor Hotel** Shipton Road, Skelton, York YO3 6XW 01904-670222 Fax 01904-670311
sB&B £69

★★★*HCR* **Grange Hotel** Clifton, York YO3 6AA 01904-644744 Fax 01904-612453
sB&B £85

★★★**Novotel** Fishergate, York YO1 4AD 01904-611660 Fax 01904-610925
sB&B £61

★★★**York Pavilion Hotel** 45 Main Street, Fulford, York YO1 4PJ 01904-622099 Fax 01904-626939
sB&B £78

★★*HR* **Heworth Court Hotel** 76-78 Heworth Green, York YO3 7TQ 01904-425156 Fax 01904-415290
sB&B £42

★★**Savages Hotel** St Peters Grove, Clifton, York YO3 6AQ 01904-610818 Fax 01904-627729
sB&B £28-£38

Derwent Lodge *Acclaimed* Low Catton, Stamford Bridge, York YO4 1EA 01759-371468
sB&B £30.50

St Georges House Hotel *Acclaimed* 6 St Georges Place, York YO2 2DR 01904-625056
sB&B £25-£30

Greenside 124 Clifton, York YO3 6BQ 01904-623631
sB&B £15-£16

The Bloomsbury Hotel 127 Clifton, York YO3 6BL 01904-634031
sB&B £10-£15

🛶 **Rawcliffe Manor Caravan Site** Manor Lane, Shipton Road, York YO3 6TZ 01904-634422

🛶 **Weir Caravan Park** Stamford Bridge, York YO4 1AN 01759-371377
Car & Caravan £7.75

Isle of Man

CASTLETOWN

★★★*HR* **Castletown Golf Links Hotel** Fort Island, Derbyhaven, Castletown 01624-822201 Fax 01624-824633
sB&B £40-£50

DOUGLAS

★★★**Sefton Hotel** Harris Promenade, Douglas 01624-626011 Fax 01624-676004
sB&B £35-£55

★★**Rutland Hotel** Queens Promenade, Douglas 01624-21218
sB&B £21.50

PORT ERIN

★★**Port Erin Imperial Hotel** Promenade, Port Erin IM9 6LH 01624-832122 Fax 01624-835402
sB&B £34.45

★★**Port Erin Royal Hotel** Promenade, Port Erin IM9 6LH 01624-833116 Fax 01624-835402
sB&B £36.45

SANTON

★★★★**Mount Murray Country Club** Santon IM4 2HT 01624-661111 Fax 01624-611116
sB&B £71.50

The Channel Islands

GUERNSEY *Guernsey*

★★★★*HCR* **La Grande Mare Hotel** The Coast Road, Vazon Bay, Castel 01481-56576 Fax 01481-56532
sB&B £74-£95

★★★**La Favorita Hotel** Fermain-Bay 01481-35666 Fax 01481-35413
In a wooded valley overlooking the bay and sea, a 3-storey, former country house with a modern wing.
& parking, ramp (or level) access, lift, wheel chair Lift, good access to restaurant, doors, adapted ground floor bedrooms, adapted en-suite bathrooms, adapted bathrooms.
sB&B £37-£45

★★★**La Trelade Hotel** Forest Road, St Martins 01481-35454 Fax 01481-37855 sB&B £28-£46.50

★★★*HCR* **St Margaret's Lodge Hotel** Forest Road, St Martins 01481-35757 Fax 01481-37594 sB&B £31-£46

La Michele *Highly Acclaimed* Les Hubits De Bas, St Martins 01481-38065 Fax 01481-39492 sB&B £20-£29

★★★★*R* **St Pierre Park Hotel** St Peter Port GY1 1FD 01481-728282 Fax 01481-712041 sB&B £95

★★★**Peninsula Hotel** Les Dicqs, Vale 01481-48400 Fax 01481-48706 sB&B £27.50-£48

JERSEY *Jersey*
★★★★*HCR* **L'Horizon** St Brelade's Bay JE3 8EF 01534-443101 Fax 01534-446269 sB&B £70-£80

★★★*HR* **Chateau de la Valeuse** St Brelade's Bay 01534-446281 Fax 01534-447110 sB&B £24-£42

★★★★*C* **Grand Hotel** Esplanade, St Helier JE4 8WD 01534-422301 Fax 01534-437815 sB&B £60

★★★*C* **Beaufort Hotel** Green Street, St Helier 01534-432471 Fax 01534-420371 sB&B £56-£69.50

★★*C* **Mont Millais Hotel** Mont Millais Street, St Helier 01534-430281 Fax 01534-466849 sB&B £27-£35

Scotland

ABERDEEN *Grampian*
★★★★**Ardoe House Hotel** Blairs, South Deeside Road, Aberdeen AB1 5YP 01224-867355 Fax 01224-861283 sB&B £88.50

★★★★**Caledonian** Union Terrace, Aberdeen AB9 1HE 01224-640233 Fax 01224-641627 sB&B £108

★★★★**Skean Dhu Hotel** Souterhead Road, Altens, Aberdeen AB1 4LF 01224-877000 Fax 01224-896964 sB&B £91.95

★★★**Dyce Skean Dhu Hotel** Farburn Terrace, Aberdeen AB2 0DW 01224-723101 Fax 01224-722965 sB&B £73

★★★**Westhill Hotel** Westhill, Aberdeen AB32 6TT 01224-740388 Fax 01224-744354 sB&B £58

ABERDEEN AIRPORT *Grampian*
★★★★**Skean Dhu Hotel** Argyll Road, Aberdeen Airport AB2 0DU 01224-725252 Fax 01224-723745 sB&B £96

ABERFELDY *Tayside*
★★*H* **Weem Hotel** Weem, Aberfeldy PH15 2LD 01887-820381 Fax 01887-820187 sB&B £24-£34

🛆🚐 **Kenmore Caravan & Camping Park** Kenmore, Aberfeldy PH15 2HN 01887-830226 Fax 01887-830211 Car & Caravan £5.50

ABERLADY *Lothian*
★★**Kilspindie House Hotel** Main Street, Aberlady EH32 0RE 01875-870687 Fax 01875-587504
♿ parking, adapted ground floor bedrooms, adapted en-suite bathrooms, sB&B £30-£42

ACHARACLE *Highland*
🛆🚐 **Loch Sunart** Resipole, Acharacle 01967-431617 Fax 01967-431777 Car & Caravan £7.50

AIRTH *Central*
ᚱ ★★★★**Airth Castle Hotel** Airth FK2 8JF 01324-831411 Fax 01324-831419 sB&B £82

ANSTRUTHER *Fife*
★★★**Craw's Nest Hotel** Bankwell Road, Anstruther KY10 3DA 01333-310691 Fax 01333-312216 sB&B £25-£40

APPIN *Strathclyde*
★★**Holly Tree** Kentallen, Appin PA38 4BY 01631-74292 sB&B £41.50

ARBROATH *Tayside*
Scurdy *Acclaimed* 33, Marketgate, Arbroath DD11 1AU 01241-872417 Fax 01241-872417 sB&B £16-£20

AUCHTERARDER *Perthshire*
🛆🚐 **Auchterarder Caravan Park** Auchterarder PH3 1ET 01764-663119 Car & Caravan £7

PLACES TO STAY

AYR *Strathclyde*
★★★**Station Hotel** Burns Statue Square, Ayr KA7 3AT 01292-263268 Fax 01292-262293 sB&B £64.60-£76.70

Brenalder Lodge *Highly Acclaimed* 39 Dunure Road, Doonfoot, Ayr KA7 4HR 01292-443939 sB&B £30-£38

Windsor Hotel *Highly Acclaimed* 6 Alloway Place, Ayr KA7 2AA 01292-264689 sB&B £20

▲⊕ **Skeldon Caravan Park** Hollybush, Ayr KA6 7EB 01292-56502 sB&B £7.50

BALLATER *Grampian*
★★★★*CR* **Craigendarroch Hotel** Braemar Road, Ballater AB35 5XA 0133-97-55858 Fax 0133-97-55447 sB&B £105

BALLOCH *Strathclyde*
★★**Balloch Hotel** Balloch G83 8LQ 01389-752579 Fax 01389-755604 sB&B £46

▲⊕ **Tullichewan Caravan Park** Old Luss Road, Loch Lomond, Balloch G83 8QP 01389-759475 Fax 01389-755563

BALMACARA *Highland*
★★**Balmacara Hotel** Balmacara IV40 8DH 0159-986-283 Fax 0159-986-329 sB&B £30-£39

BANCHORY *Grampian*
ℚ ★★★*HCR* **Raemoir Hotel** Banchory AB3 4ED 013308-24884 Fax 013308-22171 sB&B £52.30

BEAULY *Highland*
★★★**Priory Hotel** The Square, Beauly IV4 7BX 01463-782309 Fax 01463-782531 sB&B £39.75

BLAIRGOWRIE *Tayside*
★★**Angus Hotel** Blairgowrie PH10 6NQ 01250-872455 Fax 01250-875165 sB&B £25-£36

▲⊕ **Nether Craig Caravan Park** Alyth, Blairgowrie PH11 8HN 01575-560204 Fax 01575-560315 Car & Caravan £9

BOAT OF GARTEN *Highland*
▲⊕ **Campgrounds Of Scotland** Boat of Garten PH24 3BN 01479-831652 Fax 01479-831652 Car & Caravan £9

BRAEMAR *Grampian*
★★★**Invercauld Arms Hotel** Braemar AB35 5YR 0133-97-41605 Fax 0133-97-41428 sB&B £40-£63

BRORA *Highland*
★★★**Links Hotel** Brora KW9 6QS 01408-621225 Fax 01408-621383 sB&B £45-£50

CALLANDER *Central*
ℚ ★★★*HCR* **Roman Camp Hotel** Main Street, Callander FK17 8BG 01877-330003 Fax 01877-331533 sB&B £59-£79

Arran Lodge *Highly Acclaimed* Leny Road, Callander FK17 8AJ 01877-330976 ♿ramp (or level) access, sB&B £38.40-£53.60

CANNICH *Highland*
▲⊕ **Cannich Caravan Park** Cannich by Beauly, Cannich 01456-415364 Fax 01456-415263 Car & Caravan £7.50

CARRBRIDGE *Highland*
★★★*C* **Dalrachney Lodge Hotel** Carrbridge PH23 3AT 01479-841252 Fax 01479-841382 sB&B £30-£39

CASTLE DOUGLAS *Dumfries & Galloway*
▲⊕ **Loch Ken Holiday Park** Parton, Castle Douglas DG7 3NE 01644-470282 Fax 01644-470297 Car & Caravan £7

▲⊕ **Lochside Caravan Park** Castle Douglas DG7 1EZ 01644-502949 Car & Caravan £7.25

CLYDEBANK *Strathclyde*
★★★**Patio Hotel** 1 South Avenue, Clydebank Business Park, Clydebank G81 2RW 0141-9511133 Fax 0141-9523713 sB&B £67

CONNEL *Strathclyde*
Ronebhal *Acclaimed* Connel PA37 1PJ 01631-71310 sB&B £15.50-£21

CONTIN *Highland*
Coul House Hotel *Highly Acclaimed* Contin IV14 9EY 01997-421487 Fax 01997-421445 sB&B £45-£56

CRAIGELLACHIE *Grampian*
▲⊕ **Aberlour Gardens Caravan Park** Aberlour-on-Spey, Craigellachie AB38 9LP 01340-871586 Car & Caravan £5

CRAWFORD *Strathclyde*
▲⊕ **Crawford Caravan & Camp Site** Murray

Place, Carlisle Road, Crawford ML12 6TW
018642-258
Car & Caravan £6

CREETOWN *Dumfries & Galloway*
▲⊕ Creetown Caravan Park Silver Street, Creetown DG8 7HU 01671-820377
Car & Caravan £9

CRIEFF *Tayside*
★★**Locke's Acre Hotel** Comrie Road, Crieff PH7 4BP 01764-652526 Fax 01764-652526
sB&B £23-£27

★★*HC* **Murray Park Hotel** Connaught Terrace, Crieff PH7 3DJ 01764-653731 Fax 01764-675311
sB&B £45-£49

CROCKETFORD *Dumfries & Galloway*
▲⊕ Park Of Brandedleys Near Dumfries, Crocketford DG2 8RG 0155-6690250
Fax 0155-6690681
Car & Caravan £8

DINGWALL *Highland*
★★**National Hotel** High Street, Dingwall IV15 9HA 01349-62166 Fax 01349-65178
sB&B £32

DORNOCH *Highland*
▲⊕ Grannie's Heilan' Hame Embo, Dornoch IV25 3QP 01862-810383
Car & Caravan £13.50

DRYMEN *Central*
★★★**Buchanan Arms** Main Street, Drymen G63 0BQ 01360-660588 Fax 01360-660943
sB&B £75

DULNAIN BRIDGE *Highland*
★★★*HR* **Muckrach Lodge Hotel** Dulnain Bridge PH26 3LY 01479-851257 Fax 01479-851325
sB&B £43

DUMFRIES *Dumfries & Galloway*
⊕ ★★★**Hetland Hall Hotel** Carrutherstown, Dumfries DG1 4JX 01387-84201 Fax 01387-84211
sB&B £62-£75

DUNDEE *Tayside*
★★★**Queen's Hotel** Nethergate, Dundee DD1 4DU 01382-322515 Fax 01382-202668
sB&B £57-£60

★★★**Swallow Hotel** Kingsway West, Dundee DD2 5JT 01382-641122 Fax 01382-568340
sB&B £85

DUNFERMLINE *Fife*
★★★**Pitbauchlie House Hotel** 47 Aberdour Road, Dunfermline KY11 4PB 01383-722282
Fax 01383-620738
sB&B £53

★★★**Pitfirrane Arms Hotel** Main Street, Crossford, Dunfermline KY12 8NJ 01383-736132
Fax 01383-621760
sB&B £39

DUNKELD *Tayside*
▲⊕ Erigmore House Holiday Park Birnam, Dunkeld PH8 9XX 01350-727236
Fax 01350-728636
Car & Caravan £15

EAST KILBRIDE *Strathclyde*
★★★★*R* **West Point Hotel** Stewartfield Way, Phillipshill, East Kilbride G75 5LA 01355-236300
Fax 01355-233552
sB&B £95

EDINBURGH *Lothian*
★★★★★**Balmoral Hotel** 1 Princes Street, Edinburgh EH2 2EQ 0131-556-2414
Fax 0131-557-3747
sB&B £127.95-£137.95

★★★★★**Caledonian Hotel** Princes Street, Edinburgh EH1 2AB 0131-225-2433
Fax 01-03-1-225-6632
sB&B £163.50-£164.50

★★★★**Carlton Highland Hotel** North Bridge, Edinburgh EH1 1SD 0131-556-7277
Fax 0131-556-2691
sB&B £99

★★★★**Swallow Royal Scot Hotel** 111 Glasgow Road, Edinburgh EH12 8NF 0131-334-9191
Fax 0131-316-4507
sB&B £98

★★★**Capital Moat House** Clermiston Road, Edinburgh EH12 6UG 0131-334-3391
Fax 0131-334-9712
sB&B £96-£107.50

★★★**Commodore Hotel** Cramond Foreshore, Edinburgh EH4 5EP 0131-336-1700
Fax 0131-3364934
sB&B £61.75

★★★**Holiday Inn Garden Court Hotel** Queensferry Road, Edinburgh EH4 3HL 0131-3322442 Fax 0131-3323408
sB&B £81.45-£91.45

★★★*HR* **Norton House Hotel** Ingliston, Edinburgh EH28 8LX 0131-333-1275
Fax 0131-333-5305
sB&B £52-£102

★★**Iona Hotel** 17 Strathearn Place, Edinburgh EH9 2AL 0131-4475050 Fax 0131-4528574
sB&B £23-£29.50

PLACES TO STAY

Kew *Acclaimed* 1 Kew Terrace, Murrayfield, Edinburgh EH12 5JE 0131-3130700
sB&B £22-£25

Kariba Hotel 10 Granville Terrace, Edinburgh EH10 4PQ 0131-229-3773
sB&B £20-£25

ELGIN *Grampian*
★★★*R* **Mansefield House Hotel** Mayne Road, Elgin IV30 1NY 01343-540883 Fax 01343-552491
sB&B £55

FALKIRK *Central*
★★**Friendly Stop Inn** Manor Street, Falkirk FK1 1NT 01324-624066 Fax 01324-611785
sB&B £44.50-£63.50

FORFAR *Tayside*
▲⚘ **Drumshademuir Caravan Park** Roundyhill, Forfar DD8 1QT 01575-573284
Car & Caravan £7.50

FORRES *Grampian*
★★*H* **Ramnee Hotel** Victoria Road, Forres IV36 0BN 01309-672410 Fax 01309-673392
sB&B £47.50

FORT AUGUSTUS *Highland*
▲⚘ **Augustus Camping & Caravan Park** Market Hill, Fort Augustus PH32 4DH 01320-366479
Car & Caravan £6.50

FORT WILLIAM *Highland*
▲⚘ **Linnhe Caravan Park** Corpach, Fort William PH33 7NL 01397-772376
Fax 01397-772007
Car & Caravan £7.70

▲⚘ **Glen Nevis Caravan & Camping Park** Glen Nevis, Fort William PH33 6SX 01397-702191
Fax 013197-703904
Car & Caravan £6.20

GAIRLOCH *Highland*
▲⚘ **Sands Holiday Centre** Gairloch IV21 2DL 01445-712152
Car & Caravan £7.20

GATEHOUSE OF FLEET *Dumfries & Galloway*
Bank O'Fleet Hotel 47 High Street, Gatehouse of Fleet DG7 2HR 01557-814302
sB&B £19.50-£21.50

GIRVAN *Strathclyde*
★★**Westcliffe Hotel** Louisa Drive, Girvan KA26 9AH 01465-2128 Fax 01465-2128
sB&B £22-£26

GLASGOW *Strathclyde*
★★★★*R* **Moat House International** Congress Road, Glasgow G3 8QT 0141-2040733

Fax 0141-2212022
sB&B £122

★★★**Central Hotel** Gordon Street, Glasgow G1 3SF 0141-2219680 Fax 0141-2263948
sB&B £67.90-£81.10

Forte Crest Bothwell Street, Glasgow G2 7EN 0141-2212656 Fax 0141-2218986
sB&B £108.95

Marie Stuart Hotel 46-48 Queen Mary Avenue, Glasgow G42 8DT 0141-4243939
Fax 0141-4239070
sB&B £44.50

▲⚘ **Craigendmuir Park** 3 Campsie View, Stepps, Glasgow G33 6AF 0141-7794159
Fax 0141-7794057
Car & Caravan £5.50

GRANTOWN-ON-SPEY *Highland*
Ravenscourt House Hotel *Highly Acclaimed*
Seafield Avenue, Grantown-on-Spey PH26 3JG 01479-872286 Fax 01479-873260
sB&B £39-£45

HADDINGTON *Lothian*
Brown's Hotel *Highly Acclaimed* 1 West Road, Haddington EH41 3RD 01620-822254
Fax 01620-822254
sB&B £60

HAWICK *Borders*
★★**Elm House Hotel** 17 North Bridge Street, Hawick TD9 9BD 01450-72586 Fax 01450-74175
sB&B £25-£28

⚘ ★★**Mansfield House Hotel** Weensland Road, Hawick TD9 9EL 01450-73988 Fax 01450-72007
& parking, ramp (or level) access, doors, sB&B £42

HELENSBURGH *Strathclyde*
★★★**Rosslea Hall** Ferry Road, Rhu, Helensburgh G84 8NF 01436-820684
Fax 01436-820897
sB&B £63

HOWWOOD *Strathclyde*
★★★**Bowfield Hotel** Lands Of Bowfield, Howwood PA9 1DB 01505-705225
Fax 01505-705230
sB&B £65

INVERARAY *Strathclyde*
▲⚘ **Argyll Caravan & Camping Park** Inveraray PA32 8XT 01499-2285
Car & Caravan £9

INVERNESS *Highland*
★★★★**Kingsmills Hotel** Culcabock Road, Inverness IV2 3LP 01463-237166

114

Fax 01463-225208
sB&B £97

★★★Station 18 Academy Street, Inverness IV1 1LG 01463-231926 Fax 01463-710705 & parking, ramp (or level) access, lift, wheel chair Lift, good access to restaurant, adapted ground floor bedrooms, adapted en-suite bathrooms, sB&B £52-£62

★★Glen Mhor Hotel 10 Ness Bank, Inverness IV2 4SG 01463-234308 Fax 01463-713170 sB&B £57-£59

Brae Ness Hotel *Acclaimed* Ness Bank, Inverness IV2 4SF 01463-712266 sB&B £30-£33

IRVINE *Strathclyde*
★★★★Hospitality Inn Roseholme, Annick Water, Irvine KA11 4LD 01294-274272 Fax 01294-277287 sB&B £80.75

★★Redburn Hotel 65 Kilwinning Road, Irvine KA12 8SU 01294-276792 Fax 01294-76651 sB&B £24-£29

ISLE OF LEWIS
▲⊕ Laxdale Holiday Park 6 Laxdale Lane, Stornoway PA86 0DR 01851-703234

ISLE OF SKYE
★★Uig Hotel Uig IV51 9YE 01470-42205 Fax 01470-42308 sB&B £30-£45

▲⊕ Staffin Caravan & Camping Site Grenicle, Staffin IV51 9JX 0147-062-213 Car & Caravan £6

JOHN O'GROATS *Highland*
▲⊕ John O'Groats Caravan Site John O'Groats KW1 4YS 01955-611329 Car & Caravan £11

KELSO *Borders*
★★★Cross Keys 36-37 The Square, Kelso TD5 7HL 01573-223303 Fax 01573-225792 &ramp (or level) access, wheel chair Lift, sB&B £36-£42

KINGUSSIE *Highland*
★★Royal Hotel 29 High Street, Kingussie PH21 1HX 01540-661898 Fax 01540-661361 sB&B £20-£32

KINROSS *Tayside*
★★★*H* Windlestrae Hotel Kinross KY13 7AS 01577-863217 Fax 01577-864733 sB&B £65-£68

KIRKCALDY *Fife*
▲⊕ Dunnikier Caravan Park Kirkcaldy KY1 3ND 01592-267563
&Level caravan access, tent pitches with flat level access, tarmac approach, disabled toilet facilities, wheelchair access to reception/office, Car & Caravan £9

KIRKNEWTON *Lothian*
★★★★Dalmahoy Hotel Kirknewton EH27 8EB 0131-333-4092 Fax 0131-335-3203 sB&B £114

KYLE OF LOCHALSH *Highland*
★★★Lochalsh Hotel Ferry Road, Kyle of Lochalsh IV40 8AF 01599-4202 Fax 01599-4881 sB&B £55-£80

LAURENCEKIRK *Grampian*
▲⊕ Dovecot Caravan Park Northwaterbridge, Laurencekirk AB30 1QL 01674-840360 Car & Caravan £6

LOCHGILPHEAD *Strathclyde*
▲⊕ Lochgilphead Caravan Park Bank Park, Lochgilphead PA31 8NE 01546-602003 Fax 01546-603699 Car & Caravan £5

LOCKERBIE *Dumfries & Galloway*
⊕ ★★★Dryfesdale Hotel Lockerbie DG11 2SF 01576-202427 Fax 01576-204187 sB&B £47

▲⊕ Bruces Cave Kirkpatrick Fleming, Lockerbie DG11 3AT 01461-800285

▲⊕ Hoddom Castle Caravan Park Hoddom, Lockerbie DG11 1AS 01576-300251 Fax 01576-300757 Car & Caravan £10

MARKINCH *Fife*
⊕ ★★★★*HR* Balbirnie House Hotel Balbirnie Park, Markinch KY7 6NE 01592-610066 Fax 01592-610529 sB&B £85

MEY *Highland*
★★Castle Arms Hotel Mey KW14 2XH 01847-85244 Fax 01847-85244 sB&B £35

MOFFAT *Central*
★★★*HC* Moffat House Hotel High Street, Moffat DG10 9HL 01683-20039 Fax 01683-21288 sB&B £30-£52

Arden House *Acclaimed* High Street, Moffat DG10 9HG 01683-20220 sB&B £18.50

MOODIESBURN *Strathclyde*
★★★Moodiesburn Hotel 6 Cumberland Road, Moodiesburn G69 0AA 01236-873172

Fax 01236-872715
sB&B £50

MUSSELBURGH Lothian
▲⇱ **Drum Mohr Caravan Park** Levenhall, Musselburgh EH21 8JS 0131-6656867 Fax 0131-6536859
Car & Caravan £9.50

NAIRN Highland
▲⇱ **Delnies Wood Touring Park** Nairn IV12 5NX 01667-455281 Fax 01667-455437

NEW ABBEY Dumfries & Galloway
Cavens Highly Acclaimed Kirkbean, New Abbey DG2 8AA 0138-788234
sB&B £27-£35

NEWTON STEWART
▲⇱ **Cock Inn Caravan Park** Auchenmalg, Newton Stewart, DG8 0JT 01581-500227

NORTH BERWICK Lothian
★★★**Marine Hotel** Cromwell Road, North Berwick EH39 4LZ 01620-892406
Fax 01620-894480
sB&B £53-£63

OBAN Argyll
★★★**Alexandra Hotel** Corran Esplanade, Oban PA34 5AA 01631-62381 Fax 01631-64497
sB&B £35-£55

ONICH Highland
★★★*H* **Lodge on the Loch Hotel** Creag Dhu, Nr Fort William, Onich PH33 6RY 0185-5821237
Fax 0185-5821463
& parking, ramp (or level) access, adapted ground floor bedrooms, adapted en-suite bathrooms, sB&B £33-£42.50

★★*R* **Allt-nan-ros Hotel** Onich PH33 6RY 01855-3210 Fax 01855-3462
sB&B £35.50-£47.50

Tigh-a-righ Onich PH33 6SE 01855-3255
sB&B £14

PERTH Tayside
★★★*R* **Newton House Hotel** Glencarse, Perth PH2 7LX 01738-860250 Fax 01738-860717
sB&B £48-£55

★★★**Station Hotel** Leonard Street, Perth PH2 8HE 01738-624141 Fax 01738-639912
sB&B £64.60-£76.70

▲⇱ **Cleeve Caravan Park** Glasgow Road, Perth PH2 0PH 01738-39521 Fax 01738-441690

PITLOCHRY Tayside
★★★**Atholl Palace Hotel** Atholl Road, Pitlochry PH16 5LY 01796-472400 Fax 01796-473036
sB&B £35-£65

▲⇱ **Blair Castle Caravan Park** Blair Atholl, Pitlochry PH18 5SR 01796-481263
Fax 01796-481587
Top quality caravan park set amongst spectacular mountain scenery.
&Level caravan access, tent pitches with flat level access, tarmac approach, disabled toilet facilities, wheelchair access to reception/office, Specially adapted static caravans Car & Caravan £9

PORTPATRICK Dumfries & Galloway
★★★**Fernhill Hotel** Heugh Road, Portpatrick DG9 8TD 01776-810220 Fax 01776-810596
sB&B £50-£75

RHU Strathclyde
★★**Ardencaple** Shore Road, Rhu G84 8LA 01436-820200 Fax 01436-821099
sB&B £45-£51.50

ROCKCLIFFE Dumfries & Galloway
ஷ ★★★**Baron's Craig Hotel** Rockcliffe DG5 4QF 01556-630225 Fax 01556-630328
sB&B £52

ஷ ★★★*C* **Clonyard House Hotel** Colvend, Rockcliffe DG5 4QW 01556-630372
Fax 01556-630422
sB&B £30-£35

ROSYTH Fife
★★**Gladyer Inn** 10 Heath Road, Rosyth KY11 2BT 01383-419977
sB&B £32

SANDHEAD Dumfries & Galloway
▲⇱ **Sandhead Caravan Park** Sandhead 01776-830296
Car & Caravan £7.50

SELKIRK Borders
▲⇱ **Victoria Park Caravan Site** Victoria Park, Buccleugh Road, Selkirk TD7 5DN 01750-20987
Fax 01896-757003
Car & Caravan £6

SHETLAND ISLANDS
★★★**Shetland Hotel** Holmsgarth Road, Lerwick ZE1 0PW 01595-695515 Fax 01595-695828
sB&B £71

ST ANDREWS Fife
▲⇱ **Clayton Caravan Park** St Andrews KY16 9YA 01334-870242 Fax 01334-870057
Car & Caravan £11

ST BOSWELLS Borders
ஷ ★★★**Dryburgh Abbey Hotel** St Boswells TD6 0RQ 01835-822261 Fax 01835-823945
sB&B £45-£60

STEPPS *Strathclyde*
★★★ *C* **Garfield House** Cumbernauld Road, Stepps G33 6HW 0141-7792111
sB&B £60

STIRLING *Central*
★★★★*HR* **Stirling Highland Hotel** Spittal Street, Stirling FK8 1DU 01786-475444
Fax 01786-462929
Completely rebuilt hotel of character, situated close to Stirling Castle.
♿ parking, ramp (or level) access, adapted en-suite bathrooms, sB&B £85

STRACHUR *Strathclyde*
★★★(Inn) *HR* **Creggans Inn** Strachur PA27 8BX 01369-86279 Fax 01369-86637
sB&B £34-£49

STRANRAER *Dumfries & Galloway*
★★★★*H* **North West Castle Hotel** Cairnryan Road, Stranraer DG9 8EH 01776-4413
Fax 01776-2646
sB&B £49-£64

⚡ **Sands Of Luce Caravan Park** Sandhead, Stranraer DG9 9JR 01776-830456

TAIN *Highland*
★★★**Royal Hotel** High Street, Tain IV19 1AB 01862-2013 Fax 01862-3450
sB&B £40

TARBERT *Strathclyde*
⚡ **Loch Lomond Holiday Park** Inveruglas, Tarbert G83 7DW 013014-224 Fax 013014-224
Car & Caravan £10.50

TAYNUILT *Strathclyde*
⚡ **Crunachy Caravan & Camping Site** Bridge of Awe, Taynuilt PA35 1HT 018662-612
Car & Caravan £5

THORNHILL *Dumfries & Galloway*
★★**The Buccleuch & Queensberry Hotel** Thornhill 01848-330215 Fax 01848-330215
sB&B £30

THURSO *Highland*
⚡ **Thurso Caravan & Camping Site** Glebe Park, Scrabster Road, Thurso KW14 7JY 01955-603261 Fax 01955-602481
Car & Caravan £7

TONGUE *Highland*
★★**Ben Loyal Hotel** Tongue IV27 4XE 01847-55216 Fax 01847-55216
sB&B £23.50-£30

TORRIDON *Highland*
♿ ★★★**Loch Torridon** Torridon by Achnasheen IV22 2EY 01445-791242 Fax 01445-791296
sB&B £45

TROON *Ayrshire*
★★★★*HR* **Marine Highland Hotel** 8 Crosbie Road, Troon KA10 6HE 01292-314444
Fax 01292-316922
♿ parking, ramp (or level) access, lift, wheel chair Lift, good access to restaurant, doors, adapted en-suite bathrooms, adapted bathrooms, sB&B £88

TURNBERRY *Strathclyde*
★★★★★**Turnberry Hotel** Maidens Road, Turnberry KA26 9LT 01655-31000
Fax 01655-31706
sB&B £130-£170

TURRIFF *Grampian*
⚡ **Turriff Caravan Park** Station Road, Turriff 01888-62205
Car & Caravan £6

ULLAPOOL *Highland*
⚡ **Broomfield Holiday Park** Shore Street, Ullapool IV26 2SX 016854-612020
Car & Caravan £8.50

Cock Inn Caravan Park

Auchenmalg, Newton Stewart

"Get away from it all", crowds, traffic, noise peaceful select caravan park, situated on Luce Bay, on A747, coastal road between Glenluce and Fort William. Adjacent pleasant little beach and small country Inn. Panoramic views across Luce Bay to the Mull of Galloway and Isle of Man. Sailing, bathing, sea angling. (Fishing, golf and pony trekking nearby). Modern toilet block with showers and laundry room. Shop on site.

**HOLIDAY CARAVANS FOR HIRE
TOURERS WELCOME**

SAE FOR BROCHURE OR TELEPHONE 01581 500227

Wales

ABERCRAF *West Glamorgan*
Maes y Gwernen *Acclaimed* School Road, Abercrave SA9 1XD 01639-730218 Fax 01639-730765

Large house in 3/4 acre of grounds, surrounded by well kept lawns and flower beds. Encircled by walls, driveway and large car park.
& parking, ramp (or level) access, adapted en-suite bathrooms, sB&B £23

ABERDYFI (ABERDOVY) *Gwynedd*
★★★**Trefeddian Hotel** Aberdovey LL35 0SB 01654-767213 Fax 01654-767777
sB&B £28-£30

ABERPORTH *Dyfed*
_F_★★*HC* **Penbontbren Farm Hotel** Glynarthen, Cardigan, Aberporth SA44 6PE 01239-810248
sB&B £35-£40

ABERSOCH *Gwynedd*
⚘ ★★★*HR* **Porth Tocyn Hotel** Abersoch LL53 7BU 0175-881-3303 Fax 0175-881-3538
sB&B £41.50-£54

ABERYSTWYTH *Dyfed*
⛺ **Glan-Y-Mor Leisure Park** Clarach Bay, Aberystwyth SY23 3DT 01283-840640 Fax 01283-840940
Car & Caravan £9

BALA *Gwynedd*
⛺ **Bryn Melyn Leisure Park** Llandderfel, Bala LL23 7RA 01678-530121
Car & Caravan £9

⛺ **Penybont Touring & Camping Park** Llangynog Road, Bala LL23 7PH 01678-520006
Car & Caravan £5.95

BANGOR *Gwynedd*
★★★*HR* **Menai Court Hotel** Craig-y-Don Road, Bangor LL57 2BG 01248-354200

Fax 01248-354200
sB&B £47

⛺ **Treborth Hall Farm Camping & Caravan Site** Trebroth Road, Bangor LL57 2RX 01248-364399
Car & Caravan £6

BARRY *South Glamorgan*
⛺ **Fontygary Bay Caravan Park** Rhoose, Barry CF6 9ZT 01446-711074 Fax 01446-710613
Car & Caravan £10

⛺ **Vale Touring Caravan Park** Port Road (West), Barry 01446-736604
Car & Caravan £7

BEAUMARIS *Gwynedd*
★★★**Bulkeley Arms Hotel** Castle Street, Beaumaris LL58 8AW 01248-810415 Fax 01625-879335
sB&B £42-£55

BEDDGELERT *Gwynedd*
★★★**Royal Goat Hotel** Beddgelert LL55 4YE 0176-686-224 Fax 0176-686-422
sB&B £42

BETWS-Y-COED *Gwynedd*
⚘ ★★★**Plas Hall Hotel** Pont-y-pant, Betws-y-Coed LL25 0PJ 016906-206 Fax 016906-526
& parking, ramp (or level) access, doors, adapted ground floor bedrooms, adapted en-suite bathrooms, adapted bathrooms, sB&B £25-£35

★★*HR* **Ty Gwyn Hotel** Betws-y-Coed LL24 0SG 01690-710383
sB&B £19

Henllys *Acclaimed* Old Church Road, Betws-y-Coed LL24 0AL 01690-710-534
sB&B £21.50-£27

BETWS-YN-RHOS *Clwyd*
⛺ **Hunters Hamlet Touring Caravan Park** Sirior Goch Farm, Betws-yn-Rhos LL22 8PL 01745-832237

BORTH *Dyfed*
⛺ **Cambrian Coast Caravan Park** Ynyslas, Borth SY24 5JU 01970-871233 Fax 01970-871856
Car & Caravan £9

BRECON *Powys*
★★**Castle of Brecon Hotel** The Castle Square, Brecon LD3 9DB 01874-624611 Fax 01874-623737
sB&B £49

PLACES TO STAY

🅰️🚙 **Brynich Caravan Park** Brecon LD3 7SH
01874-623325 Fax 01874-623325
Car & Caravan £7.20

BRONLLYS *Powys*
🅰️🚙 **Anchorage Caravan Park** near Brecon,
Bronllys LD3 0LD 01874-711246
Car & Caravan £6

BRYNTEG *Gwynedd*
🅰️🚙 **Nant Newydd Caravan Park** Brynteg LL78
8JH 01248-852842
Car & Caravan £7

BUILTH WELLS *Powys*
❊ ★★**Pencerrig Gardens Hotel** Builth Wells LD2
3TF 01982-553226 Fax 01982-552347

sB&B £43.50-£47.50

CAERNARFON *Gwynedd*
🅰️🚙 **Bryn Gloch Caravan & Camping Park**
Betws Garmon, Caernarfon LL54 7YY
01286-650216 Fax 01286-650216
Car & Caravan £7.40

🅰️🚙 **Morfa Lodge Caravan Park** Dinas Dinlle,
Caernarfon LL54 5TP 01286-830205
Fax 01286-831329
Car & Caravan £8.50

CARDIFF *South Glamorgan*
★★★★**Cardiff International Hotel** Mary Ann
Street, Cardiff CF1 2EQ 01222-341441
Fax 01222-223742
♿ parking, ramp (or level) access, wheel chair
Lift, adapted en-suite bathrooms, sB&B £83.95-
£93.95

★★★★**Copthorne Hotel** Copthorne Way,
Culverhouse Cross, Cardiff CF5 6XJ
01222-599100 Fax 01222-599080
sB&B £101.95

★★★**Churchills Hotel** Llandaff Place, Cardiff
Road, Llandaff, Cardiff CF5 2AD 01222-562372
Fax 01222-568147
sB&B £66.90

★★★**St Mellons Country Club** St Mellons,
Cardiff CF3 8XR 01633-680355 Fax 01633-680399
sB&B £45

Forte Crest Castle Street, Cardiff CF1 2XB
01222-388681 Fax 01222-371495
sB&B £85.50

Pavilion Lodge Pontclun, Cardiff CF7 8SB
01222-892253 Fax 01222-892497
sB&B £37.45

COLWYN BAY *Clwyd*
★★**Ashmount Hotel** College Avenue, Rhos-on-
Sea, Colwyn Bay LL28 4NT 01492-544582
Fax 01492-545479
sB&B £29.75-£37

★★**R Hopeside Hotel** 63 Princes Drive, Colwyn
Bay LL29 8PW 01492-533244 Fax 01492-532850
sB&B £35

Northwood Hotel 47 Rhos Road, Rhos-on-Sea,
Colwyn Bay LL28 4RS 01492-549931
sB&B £18

CONWY *Gwynedd*
❊ ★★★**Sychnant Pass Hotel** Sychnant Pass
Road, Conwy LL32 8BJ 01492-596868
Fax 01492-870009
sB&B £40

CRICCIETH *Gwynedd*
★★**Gwyndy Hotel** Llanystumdwy, Criccieth LL52
0SP 01766-522720
sB&B £27

CRICKHOWELL *Powys*
🅰️🚙 **Riverside Caravan & Camping Park** New
Road, Crickhowell NP8 1AY 01873-810397
Fax 01873-811989
Car & Caravan £6

CWMBRAN *Gwent*
★★★★**Parkway Hotel** Cwmbrian Drive,
Cwmbran NP44 3UW 01633-871199
Fax 01633-869160
sB&B £78.45

DOLGELLAU *Gwynedd*
★★**(Inn) George III Hotel** Penmaenpool,
Dolgellau LL40 1YD 01341-422525
Fax 01341-423565
♿ parking, ramp (or level) access, doors, sB&B
£45

★**Clifton House Hotel** Smithfield Square,
Dolgellau LL40 1ES 01341-422554
sB&B £25.50-£32.50

🚲 **R Fronoleu Farm Hotel** Tabor, Dolgellau
LL40 2PS 01341-422361
sB&B £17.50-£23.50

DYFFRYN VALLEY *Gwynedd*
★★**Bull Hotel** London Road, Dyffryn Valley
LL65 3DP 01407-740351 Fax 01407-742328
sB&B £29.75

EWLOE *Clwyd*
★★★★**St David's Park** St David's Park, Ewloe
CH5 3YB 01244-520800 Fax 01244-520930
sB&B £87.95

FISHGUARD *Dyfed*
⚠️🏕️ **Gwaun Vale Holiday Touring Park**
Llanychaer, Fishguard SA65 9TA 01348-874698
Car & Caravan £8.50

GWBERT-ON-SEA *Dyfed*
★★★**Cliff Hotel** Gwbert-on-Sea SA43 1PP
01239-613241 Fax 01239-615391
sB&B £25-£28.50

LLANDOVERY *Dyfed*
⚠️🏕️ **Erwlon Caravan & Camping Park**
Llandovery SA20 0RD 01550-20332
Car & Caravan £5

LLANDUDNO *Gwynedd*
★★**Esplanade Hotel** Central Promenade,
Llandudno LL30 2LL 01492-860300
Fax 01492-860418
sB&B £24-£36

★★**Royal Hotel** Church Walks, Llandudno LL30
2HW 01492-876476 Fax 01492-870210
sB&B £22.50

★*C* **Epperstone Hotel** 15 Abbey Road,
Llandudno LL30 2EE 01492-878746
Fax 01492-871223
sB&B £21-£24

LLANDWROG *Gwynedd*
⚠️🏕️ **White Tower Caravan Park** Llandwrog LL54
5UH 01286-830649
Car & Caravan £5.70

LLANGEFNI *Gwynedd*
🍴 ★★★*HCR* **Tre-ysgawen Hall Hotel** Capel
Coch, Llangefni LL77 7UR 01248-750750
Fax 01248-750035
sB&B £78.50

MACHYNLLETH *Brecon*
★★**White Lion** Heol Pentrenhedyn, Machynlleth
SY20 8ND 01654-703455 Fax 01654-703746
sB&B £20-£30

⚠️🏕️ **Warren Parc Caravan Park** Penegoes,
Machynlleth SY20 8NN 01654-702054
Car & Caravan £8

MERTHYR TYDFIL *Mid Glamorgan*
★★*HC* **Tregenna Hotel** Park Terrace, Merthyr
Tydfil CF47 8RF 01685-723627 Fax 01685-721951
sB&B £39

MOLD *Clwyd*
★★★**Beaufort Palace** Bryn Offa Lane, New
Brighton, Mold CH7 6RQ 01352-758646
Fax 01352-757132
sB&B £60

MUMBLES *West Glamorgan*
★**St Anne's Hotel** Western Lane, Mumbles SA3
4EY 01792-369147 Fax 01792-360537
sB&B £30-£33

NEW QUAY *Dyfed*
Brynafor Hotel *Acclaimed* New Road, New Quay
SA45 9SB 01545-560358
sB&B £25-£30

🏡 **Ty Hen Farm** *Acclaimed* Llwyndafydd, New
Quay SA44 6BZ 01545-560346
sB&B £20-£29

NEWCASTLE EMLYN *Dyfed*
⚠️🏕️ **Afon Teifi Caravan & Camping Park** Pentre
Cagal, Newcastle Emlyn SA38 9HT 01559-370532
Car & Caravan £6.50

NEWPORT *Gwent*
★★★**Westgate Hotel** Commercial Street,
Newport NP9 1TT 01633-244444
Fax 01633-246616
sB&B £64

PORT TALBOT *West Glamorgan*
★★★**Aberavon Beach Hotel** Port Talbot SA12
6QP 01639-884949 Fax 01639-897885
sB&B £53

⚠️🏕️ **Afan Argoed Country & Touring Park**
Cynonville, Port Talbot SA13 3HG 01639-850564
Fax 01639-895897

SAUNDERSFOOT *Dyfed*
★★**Rhodewood House Hotel** St Brides Hill,
Saundersfoot SA69 9NU 01834-812200
Fax 01834-811863
sB&B £23-£30

Jalna Hotel *Acclaimed* Stammers Road,
Saundersfoot SA69 9HH 01834-812282
sB&B £21-£23

🏡 **Vine Farm** *Acclaimed* The Ridgeway,
Saundersfoot SA69 9LA 01834-813543
sB&B £20

SOLVA *Dyfed*
🏡 **Lochmeyler Farm** *Highly Acclaimed* Pen-y-
Cwm, Llandeloy, Solva, Haverfordwest SA62 6LL

01348-837724 Fax 01348-837724
sB&B £20-£30

ST ASAPH *Clywd*
★★★**Talardy Park Hotel** The Roe, St Asaph
01745-584957 Fax 01745-584385
Once a country manor house, the hotel retains an air of grandeur and is set in extensive grounds. High standard of decoration throughout.
&ramp (or level) access, sB&B £41.50

ST DAVID'S *Dyfed*
Å🏕 **Tretio Caravan & Camping Park** St David's
SA62 6DE 01437-720270
Car & Caravan £6.50

SWANSEA *West Glamorgan*
★★★**Dolphin Hotel** Whitewalls, Swansea SA1
3AB 01792-650011 Fax 01792-642871
sB&B £49.50-£59.50

Å🏕 **Riverside Caravan Park** Ynysforgan Farm,
Morriston, Swansea SA6 6QL 01792-775587
Car & Caravan £9

TENBY *Dyfed*
★★★*HC* **Atlantic Hotel** Esplanade, Tenby SA70
7DU 01834-2881 Fax 01834-2881
sB&B £45-£51

★★**Green Hills** St Florence, Tenby 01834-871291
Fax 01834-871948
sB&B £22-£25

★★**Royal Lion Hotel** High Street, Tenby SA70
7EX 01834-842127 Fax 01834-842441
sB&B £25

Waterwynch House Hotel *Highly Acclaimed*
Waterwynch Bay, Tenby SA70 8TJ 01834-842464
Fax 01834-845076
sB&B £30-£38

TOWYN *Clwyd*
Å🏕 **Ty Mawr Holiday Park** Towyn Road, Towyn
LL22 9HG 01442-230300 Fax 01442-230368
Car & Caravan £7.50

TREDEGAR *Gwent*
Å🏕 **Bryn Bach Park** Merthyr Road, Tredegar
NP2 3AY 01495-711816 Fax 01495-726630
Car & Caravan £6

WREXHAM *Clwyd*
ቧ ★★★**Llwyn Onn Hall Hotel** Cefn Road,
Wrexham LL13 0NY 01978-261225
Fax 01978-261225
sB&B £52

Northern Ireland

AGHADOWEY *Co Londonderry*
★★**(Inn) Brown Trout Golf and Country Club**
209 Agivey Road, Aghadowey BT51 4AD
01265-868209 Fax 01265-868878
sB&B £40

BELFAST *Co Antrim and Co Down*
★★★**Dukes Hotel** 65 University Street, Belfast
BT7 1LH 01232-236666 Fax 01232-237177
sB&B £81

CRAWFORDSBURN *Co Down*
★★★**Old Inn** Crawfordsburn BT19 1JH
01247-853255 Fax 01247-852775
sB&B £70

DUNGANNON *Co Tyrone*
★★★**Inn on the Park** Moy Road, Dungannon
BT71 6BS 01868-725151 Fax 01868-724953
sB&B £37.50

HOLYWOOD *Co Down*
★★★★*R* **Culloden Hotel** Craigavad, Holywood
BT18 0EX 0123-17-5223 Fax 0123-17-6777
sB&B £108

IRVINESTOWN *Co Fermanagh*
★★**Mahons Hotel** Enniskillen Road, Irvinestown
BT74 9XX 01365-621656 Fax 01365-628344
sB&B £29.50

LISNASKEA *Co Fermanagh*
Å🏕 **Share Centre** Smith's Strand, Lisnaskea
013657-22122 Fax 013657-21893
Car & Caravan £8

NEWTOWNABBEY *Co Antrim*
★★★**Chimney Corner Motel** 630 Antrim Road,
Newtownabbey BT36 8RH 01232-844925
Fax 01232-844352
sB&B £69.50

Republic of Ireland

ARKLOW *Co Wicklow*
★★**Lawless's Hotel** Aughrim, Arklow 0402-36146
Fax 0402-36384
sB&B £37-£41

BALLYMACODA *Co Cork*
▲☂ **Sonas Caravan & Camping Park**
Ballymacoda 024-98132
Car & Caravan £7

BALLYVAUGHAN *Co Clare*
☙ ꝗ ★★★**Gregans Castle Hotel**
Ballyvaughan 065-77005 Fax 065-77111
sB&B £62-£136

BANDON *Co. Cork*
▲☂ **Murrays Caravan & Camping Park**
Kilbrogan Farm., Bandon 023-41232

BANTRY *Co Cork*
★★★**Westlodge Hotel** Bantry 027-50360
Fax 027-50438
sB&B £35-£45

BELCARRA *Co. Mayo*
▲☂ **Carra Caravan And Camping Park**
Castlebar, Belcarra 094-32054
Car & Caravan £5

BENNETTSBRIDGE *Co. Kilkenny*
▲☂ **Nore Valley Park** Bennettsbridge 056-27229
Car & Caravan £7

BLARNEY *Co Cork*
★★★**Blarney Park Hotel** Blarney 021-385281
Fax 021-381506
sB&B £49-£55

BUNRATTY *Co Clare*
Bunratty View *Highly Acclaimed* Cratloe, Bunratty
061-357352
sB&B £18

CASHEL *Co. Galway*
☙ ꝗ ★★★**Cashel House Hotel**
Connemara, Cashel 095-31001 Fax 095-31077
sB&B £51.75-£70.87

Zetland House Hotel Cashel Bay, Cashel
095-31111
sB&B £57-£75

CORK *Co Cork*
★★★★**Fitzpatrick Silver Springs Hotel** Tivoli,
Cork 021-507533 Fax 021-507641
sB&B £77.50

COURTOWN *Co. Wexford*
▲☂ **Parklands Holiday Park** Ardamine,
Courtown Gorey, Courtown 055-25202

Fax 055-25202
Car & Caravan £12

DONEGAL *Co. Donegal*
★★★**HR Harvey`s Point Country Hotel** Lough
Eske, Donegal 073-22208 Fax 073-22352
sB&B £55-£60

DUBLIN *Co Dublin*
★★★★★**Berkeley Court Hotel** Lansdowne
Road, Dublin 01-6601711 Fax 01-6617238
sB&B £47-£150

★★★★★**Conrad Hotel** Earlsfoot Terrace,
Dublin 01-6765555 Fax 01-6765424
sB&B £145

★★★★**H Gresham** O'Connell Street, Dublin
01-8746881 Fax 01-8787175
sB&B £116

★★★★**Jurys Hotel and Towers** Pembroke Road,
Ballsbridge, Dublin 01-6605000 Fax 01-6605540
sB&B £105-£150

★★★**Jurys Christchurch Inn** Christchurch Pl,
Dublin 01-4750111 Fax 01-4750488
sB&B £48

★★★**Temple Bar Hotel** Dublin 01-6773333
Fax 01-6773088
sB&B £77.50-£82.50

★★**Holly Brook Hotel** Holly Brook Park,
Clontare, Dublin 01-8335456 Fax 01-8335458
sB&B £35

★★**HCR Longfields Hotel** 10 Lower Fitzwilliams
Street, Dublin 01-6761367 Fax 01-6761542
sB&B £65-£79.50

Ariel House *Highly Acclaimed* 50-52 Lansdowne
Road, Ballsbridge, Dublin 01-6685512
Fax 01-6685845
sB&B £46.50-£52.50

DUGORT *Co Mayo*
▲☂ **Keel Sandybanks Caravan & Camping Park**
Achill Island, Dugort 094-32054
Car & Caravan £7

DUNDALK *Co Louth*
★★**New Fairways Hotel** Dublin Road, Dundalk
042-21500 Fax 042-21511
sB&B £45

DUNGARVAN *Co Waterford*
▲☂ **Casey's Caravan Park** Clonea, Dungarvan
058-41919
Car & Caravan £9

Places to Stay

ENNIS *Co Clare*
★★★**Auburn Lodge Hotel** Galway Road, Ennis
065-21247 Fax 065-21202
sB&B £30

★★★**West County Inn** Clare Road, Ennis
065-28421 Fax 065-28801
sB&B £45-£54

★★**Queens Hotel** Abbey Street, Ennis 065-28963
Fax 065-28628
sB&B £30-£45

FURBO *Co Galway*
★★★★**Connemara Coast Hotel** Furbo 091-92108
Fax 091-92065
sB&B £57.50-£77

GALWAY & SALTHILL *Co Galway*
★★★★**C Glenco Abbey Hotel** Galway 091-26666
Fax 091-27800
sB&B £68-£80

★★★**Jurys Galway Inn** Quay Street, Galway
091-66444 Fax 091-68415
sB&B £46-£56

GLENGARRIFF *Co Cork*
★★**Casey's Hotel** Glengarriff 027-63072
Fax 027-63010
sB&B £27-£30

GOREY *Co Wexford*
🍽 ⊕ ★★★**Marlfield House Hotel** Gorey
053-457775 Fax 053-457779
sB&B £60-£90

HOWTH *Co Dublin*
★★★**Deer Park Hotel & Golf Courses** Howth,
Dublin 01-8322624 Fax 01-8392405
sB&B £50-£55

INNISHANNON *Co Cork*
★★**Innishannon House Hotel** Innishannon
021-775121 Fax 021-775609
sB&B £50-£65

KENMARE *Co Kerry*
🍽 ⊕ ★★★★**Park Hotel** Kenmare
064-41200 Fax 064-41402
sB&B £112-£130

🍽 ★★★★**Sheen Falls Hotel** Kenmare
064-41600 Fax 064-41386
sB&B £125-£175

KILLARNEY *Co Kerry*
★★★★**HCR Aghadoe Heights Hotel** Killarney
064-31766 Fax 064-31345
sB&B £80-£110

★★★**Castlerosse Hotel** Killarney 064-31144
Fax 064-31031
sB&B £49-£66

★★★**International** Killarney 064-31816
Fax 064 31837
sB&B £37-£47

★★★**Killarney Ryan Hotel** Killarney 064-31555
sB&B £96-£128

Glena House Muckross Road, Killarney
064-35611 Fax 064-35611
sB&B £22-£24

▲⊕ **Fossa Caravan & Camping Park** Fossa,
Killarney 064-31497 Fax 064-34459

▲⊕ **The Flesk** Muckross Road, Killarney
064-31704

▲⊕ **White Villa Farm Caravan & Camping Site**
Cork Road, Killarney 064-32456

KILLINEY BAY *Co Dublin*
★★★**The Court Hotel** Killiney Bay 01-2851622
Fax 01-2852085
sB&B £63-£81.56

KILMUCKRIDGE *Co Wexford*
▲⊕ **Morriscastle Strand Caravan Park**
Kilmuckridge 01-4535355 Fax 01-4545916
♿Level caravan access, tent pitches with flat level
access, disabled toilet facilities,
Car & Caravan £8.50

KINSALE *Co Cork*
★★★**Actons** Pier Road, Kinsale 024-772135
Fax 021-772231
sB&B £55-£70

Moorings *Highly Acclaimed* Scilly, Kinsale
021-772675 Fax 021-772675
sB&B £35-£50

LAURAGH VILLAGE *Co Kerry*
▲⊕ **Creveen Lodge Caravan & Camping Park**
Healy Pass, Lauragh Village 064-83131

LIMERICK *Co Limerick*
★★★★**Limerick Inn Hotel** Ennis Road,
Limerick 061-326666 Fax 061-326281
sB&B £84-£91

★★★★**Limerick Ryan Hotel** Ennis Road,
Limerick 061-453922 Fax 061-326333
sB&B £81-£105

★★**Woodfield House Hotel** Ennis Road,
Limerick 061-453023 Fax 061-326755
sB&B £35-£40

MACROOM *Co Cork*
Mills Inn *Highly Acclaimed* Ballyvourney,
Macroom 026-45237 Fax 026-45454
sB&B £25-£30

MAYNOOTH *Co Kildare*
⊕ ★★★**HCR Moyglare Manor Hotel** Maynooth

01-6286351
sB&B £75-£80

NAAS *Co Kildare*
★★**Harbour View Hotel** Limerick Road, Naas
045-79145 Fax 045-74002
sB&B £35

ORANMORE *Co Galway*
Moorings *Acclaimed* Main Street, Oranmore
091-90462
sB&B £20-£25

OUGHTERARD *Co Galway*
★★★*H* **Connemara Gateway Hotel** Oughterard
091-82328 Fax 091-82332
sB&B £50-£75

⊕ ★★★**Ross Lake House Hotel** Rosscahill,
Oughterard 091-80109 Fax 091-80184
sB&B £43-£48.50

RATHNEW *Co Wicklow*
⊕ ★★★**Tinakilly Country House Hotel**
Wicklow, Rathnew 0404-69274 Fax 0404-67806
sB&B £80-£86

ROSCOMMON *Co Roscommon*
★★★**Abbey Hotel** Roscommon 0242-51605
Fax 0242-51303
sB&B £40-£50

ROSSLARE *Co Wexford*
★★★★*HR* **Kelly's Strand Hotel** Rosslare
053-32114 Fax 053-32222
sB&B £41.80-£52.80

SHANKILL *Co Dublin*
▲⊕ **Shankill Caravan & Camping Park**
Sherrington Park., Shankill 01-2820011
Car & Caravan £12

SNEEM *Co Kerry*
Tahilla Cove Country House *Highly Acclaimed*
Tahilla, Sneem 064-45204
sB&B £40-£45

STRAFFAN *Co Kildare*
★★★★★**Kildare** Straffan 062-7-3333
sB&B £155-£180

VIRGINIA *Co. Cavan*
▲⊕ **Lough Ramor Caravan Park** Rycfield,
Virginia 049-47447
Car & Caravan £6

WATERFORD *Co Waterford*
★★★★*H* **Granville Hotel** Waterford 051-55111
Fax 051-70307
sB&B £45-£50

WATERVILLE *Co Waterford*
★★★**Butler Arms Hotel** Waterville 066-74144
Fax 066-74520
sB&B £65-£75

▲⊕ **Waterville Caravan & Camping Park**
Waterville 066-74191 Fax 066-74538
Car & Caravan £6.50

WEXFORD *Co Wexford*
★★★**Ferrycarrig Hotel** Ferrycarrig Bridge, P.O.
Box 11, Wexford 053-22999 Fax 053-41982
sB&B £39-£49

YOUGHAL *Co Cork*
H **Ahernes Seafood Restaurant & Hotel** 163
Main Street, Youghal 024-92424 Fax 024-93633
sB&B £35-£60

ROSEMOOR COUNTRY COTTAGES
PEMBROKESHIRE COAST NATIONAL PARK

Our warm and comfortable red sandstone cottages provide every home comfort, from wall to wall carpeting to full gas-fired central heating, in lovely countryside two miles from the sea at Little Haven. They were commended by the Countryside Commission 18 months ago as an example of good practice in a National Park. The Rosemoor Nature Reserve (managed by the Dyfed Wildlife Trust) includes a 5 acre lake, where, if you're lucky, you'll see otters at play. Five cottages are all ground-floor and are suitable for assisted wheel-chair users, and one has bed - and en-suite bathroom equipped with necessary aids.

Please send for our "exceptionally interesting brochure" (Good Holiday Cottage Guide).
JOHN & BARBARA LLOYD, ROSEMOOR, WALWYN'S CASTLE, HAVERFORDWEST SA62 3ED. Tel: 0437 781326 Fax: 0437 781080

The Country House Hotel For Disabled People

Park House, situated in beautiful surroundings on the Sandringham Royal Estate, provides superb holiday facilities throughout the year

- Award winning hotel
- All physical disabilities accommodated
- Care available at no additional cost
- Relatives and companions equally welcome

LOWER PRICES IN WINTER MONTHS

Send for brochure and tariff
Park House Sandringham, King's Lynn, Norfolk PE35 6EH
Tel: **Dersingham (0485) 543000** Fax: **(0485) 540663**

THE *COMPLETE* HOLIDAY
WHATEVER YOUR DISABILITY
The Leonard Cheshire Foundation